Nonlinear Economic Models

Nonlinear Economic Models

Cross-sectional, Time Series and Neural Network Applications

Edited by

John Creedy

Truby Williams Professor of Economics, University of Melbourne

Vance L. Martin

Associate Professor of Economics, University of Melbourne

Edward Elgar
Cheltenham, UK • Lyme, US

Published by
Edward Elgar Publishing Limited
8 Lansdown Place
Cheltenham
Glos GL50 2HU
UK

Edward Elgar Publishing, Inc.
1 Pinnacle Hill Road
Lyme
NH 03768
US

A catalogue record for this book
is available from the British Library

Library of Congress Cataloguing in Publication Data
Nonlinear economic models : cross-sectional, times series and neural
network applications / edited by John Creedy, Vance L. Martin.
Includes index
1. Econometric models. 2. Nonlinear theories. 3. Linear models
(Statistics) I. Creedy, John. 1949– . II. Martin, Vance. 1955–

HB141.N657 1997 97–23224
330'.01'5195—dc21 CIP

ISBN 1 85898 637 0
Printed and bound in Great Britain by Hartnolls Limited, Bodmin, Cornwall

Contents

II Cross-sectional Applications

List of Figures

List of Tables

Contributors

Alex Bakker is a PhD student in Economics at the University of Melbourne.

Eugene S.Y. Choo is a PhD student in Economics at Yale University.

John Creedy is the Truby Williams Professor of Economics at the University of Melbourne.

David Dickson is a Senior Lecturer in Actuarial Studies at the University of Melbourne.

G.C. Lim is an Associate Professor of Economics at the University of Melbourne.

Jenny N. Lye is a Senior Lecturer in Economics at the University of Melbourne.

Vance L. Martin is an Associate Professor in Economics at the University of Melbourne.

Gael M. Martin is a Research Fellow at Monash University.

P.D. McNelis is a Professor of Economics at Georgetown University.

Robert Pereira is a Lecturer in Economics at LaTrobe University and is a PhD student in Economics at the University of Melbourne.

Clarence Tan is a PhD student in Finance at the Bond University.

Leslie Teo is an Economist at the International Monetary Fund.

Part I

Introduction

Chapter 1

Nonlinear Modelling: An Introduction

John Creedy and Vance L. Martin

This chapter introduces nonlinear economic modelling using the generalised exponential family of distributions. The approach is very flexible and extremely powerful. The development of this type of nonlinear modelling is enhanced by advances in computers which have made possible the use of computationally intensive numerical iterative procedures. However, the chapter does not emphasise the technical details involved, particularly concerning methods of estimation. Its aim to illustrate the power of the approach, to show how it can be used in a variety of economic contexts, and to provide an introduction to the more technical chapters that follow. The method is illustrated by examining a time series of hypothetical data and comparing nonlinear and linear approaches. It is suggested that there are contexts in which the assumptions of linear models are not appropriate, even though they may sometimes appear to produce quite good results.

An economic model to be estimated is characterised by a specification in which an 'endogenous' variable is related in some way to a set of 'exogenous' variables. In linear models it is possible to express the model, using a transformation of variables if necessary, so that the conditional value of the endogenous variable is expressed as a linear function of the exogenous variables. The conditional distribution of the endogenous variable is usually regarded as following the normal distribution or one of the other standard unimodal distributions. This chapter discusses the advantages of using, instead of a given form of unimodal distribution, the family of distributions known as the generalised exponential family. Examples of the application of this family are given in later chapters; see also Cobb *et al.* (1983), Cobb and Zacks (1985), Creedy and Martin (1994b,c), Fischer and Jammernegg (1986), Lye and Martin (1993a), and Martin (1990).

Nonlinear modelling using the generalised exponential family is particularly useful because it can handle phenomena such as large discrete 'jumps' in the endogenous variable that cannot easily be modelled using modifications of the linear model, such as the addition of dummy variables which require the investigator to specify precisely when the jump takes place. This aspect is investigated in chapter 9. A very important feature is that the approach is able to provide a link between simple economic models and the distributional forms of economic variables. This link can be derived directly from the structural form of the model rather than involving the more or less arbitrary imposition of a form of distribution at the estimation stage. The flexibility of the approach means that a wide range of models can be 'nested' and therefore more easily compared. Many examples of alternative nestings are given throughout this book. Very importantly, it also allows for the possibility that the form of the conditional distributions may themselves change over time, and indeed may sometimes be multimodal.

In order to illustrate alternative approaches, a hypothetical time series of prices is used. The prices are regarded as being influenced by two fundamental variables that are treated as exogenous. Results using a standard linear approach are obtained in Section 1.1. This section also presents an alternative way of specifying the linear model, which facilitates an immediate comparison with the nonlinear approach. The extension of the linear model is presented in Section 1.2, where the generalised exponential family of distributions is introduced. The results of estimating a nonlinear model are contrasted with the use of the linear model. A further valuable feature of the approach is that, starting from a simple economic model, it is possible to generate the distributional implications of a stochastic process imposed on the model. The way in which the distributions can be generated is described in Section 1.3. Some examples showing how the general result can be applied in special cases are provided in Section 1.4.

1.1 Linear Models

Basic courses in econometrics stress the idea that in a standard regression analysis each observed value of an endogenous or dependent variable is usually regarded as being an observation selected at random from a conditional distribution, for a given set of values of exogenous or independent variables. The estimation and testing of models proceeds by specifying an assumed form for the conditional distributions, and this is combined with the idea of random independent selection. In practice, only one observation from each conditional distribution is available, so the form of the conditional distribution cannot be observed directly. The assumptions that the conditional means are given by a linear combination of the exogenous variables while the conditional variances are homoscedastic are therefore very convenient. With the additional assumption of normality of the conditional distributions, it is easy to show that maximum likelihood estimation reduces to ordinary least squares and has other desirable properties such as efficiency.

Within the context of this type of linear model, it is possible to allow for additional complexities, arising for example if the selection process is not random so that the probability of observing a value from one conditional distribution depends in some way on the values drawn from other conditional distributions, or if the conditional variances are not all the same. When such modifications to the basic model are made, such as assuming that the data generation process involves some serial correlation or introducing heteroscedasticity, the assumption is retained that all the conditional distributions take the same fundamental form, that is the unimodal normal distribution. Even where non-normal distributions are specified, it is typically assumed that they are unimodal.

This assumption of unimodality often appears to be supported by the available evidence. For example, in the context of the analysis of a stationary time series of relative prices, it is possible to produce an unconditional distribution by a simple process of 'temporal aggregation', whereby all the relative prices are treated as coming from the same distribution. Such empirical distributions obtained in this way are typically unimodal, although they tend to exhibit sharper peaks and fatter tails than a normal distribution with the same mean and variance. For examples in the context of exchange rate returns, see Boothe and Glassman (1987), Friedman and Vandersteel (1982), Hsieh (1989b), Mandelbrot (1963), and Tucker and Pond (1988). Further discussion of exchange rate models is given in chapters 8 and 14.

1.1.1 Estimation and Prediction

Suppose a set of data is available, consisting of a time series of 200 observations on a variable, p_t, that will for convenience be called a price. The price is thought to be influenced by two 'fundamental' exogenous variables, x_t and y_t. The precise model used to generate these hypothetical time series is described below; in practice the 'true' model must of course remain unknown.

Linear regression analysis proceeds by specifying an equation in which the price, or a simple transformation such as its logarithm, is a linear function of the exogenous variables, x_t and y_t, or transformed values, and an additive stochastic term, u_t. Faced with the data, alternative specifications, within

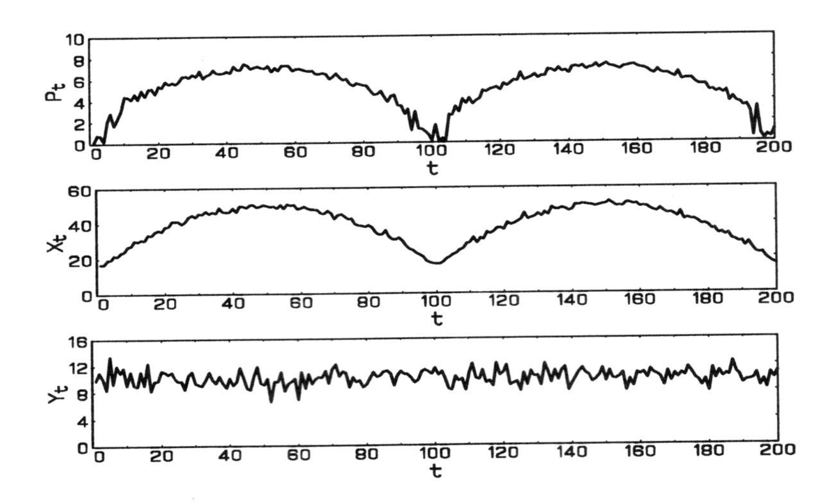

Figure 1.1: Data series

the context of linear regression analysis, would be estimated and a range of tests applied, starting with the simplest assumptions about the stochastic term, u_t.

Suppose that the linear model is specified as:

$$p_t = \alpha_0 + \alpha_1 x_t + \alpha_2 y_t + u_t \tag{1.1}$$

with the u_ts assumed to be independently distributed as $N(0, \sigma_u^2)$. Each value of u_t is regarded as being independently drawn from a distribution which is assumed to have the same form, $N(0, \sigma_u^2)$, irrespective of the values of x_t and y_t. This implies that prices are also normally distributed, although around a time-varying conditional mean which is a function of both x_t and y_t. Estimation by ordinary least squares produces the following results, where absolute values of t-statistics are given in brackets:

$$\begin{array}{lllll} \widehat{p_t} = - & 1.746 & + \quad 0.185 & x_t - & 0.020 \quad y_t \\ & (5.068) & (59.166) & & (0.663) \end{array} \tag{1.2}$$

with $\widehat{\sigma}_u = 0.462$; $\overline{R}^2 = 0.947$; $DW = 1.327$;.$BP(2) = 69.142$; $ARCH(1) = 45.018$; $RESET(1) = 68.213$.[1]

[1]The reported diagnostics are: $\widehat{\sigma}_u$, the standard error of estimate; $\overline{R}^2$, the degrees of

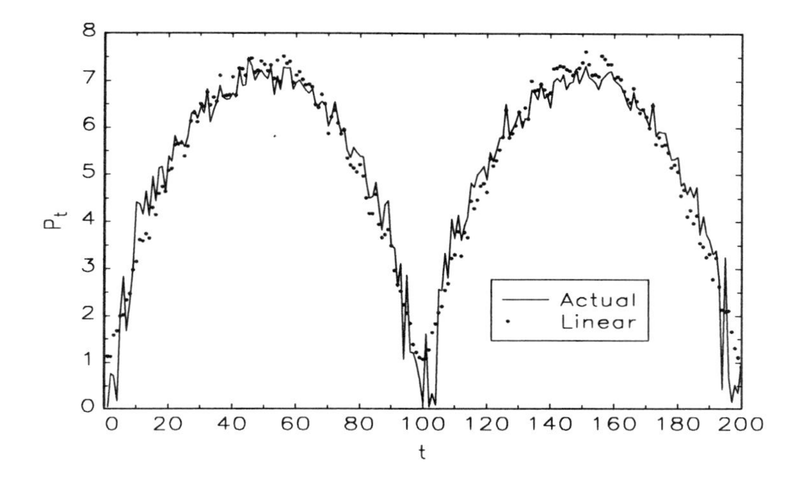

Figure 1.2: Actual and predicted p_t based on the linear model

Using conventional criteria of goodness of fit, the linear model performs well. The value for $\overline{R}^2$ shows that a high proportion of variation in p_t is explained by a linear relationship between the dependent variable and the independent variables x_t and y_t, as well as the intercept term. A comparison of the actual and the predicted price series, $\widehat{p_t}$, that is the conditional mean in each period, is shown in Figure 1.2. This shows that the linear model is able to predict very well the general swings in p_t over most of the sample. However, closer inspection of Figure 1.1 reveals some evidence of misspecification as the linear model fails to capture the large swings in prices at the start, middle and end of the sample period. Evidence of misspecification is also highlighted by the DW and BP statistics which both suggest significant serial correlation and heteroscedasticity respectively. These results suggest that there is some underlying characteristic which is not being captured by this linear model. This is supported by the $ARCH$ and $RESET$ test statistics which suggest that there is strong evidence of nonlinearities in the residuals.

freedom adjusted coefficient of determination; DW, the Durbin-Watson statistic for first order autocorrelation; $BP(2)$, the Breusch-Pagan statistic for heteroscedasticity which is distributed as χ_2^2; $ARCH(1)$, Engle's statistic for testing for first order autoregressive conditional heteroscedasticity which is distributed as χ_1^2; $RESET(1)$, statistic to test for nonlinearities which is distributed as χ_1^2.

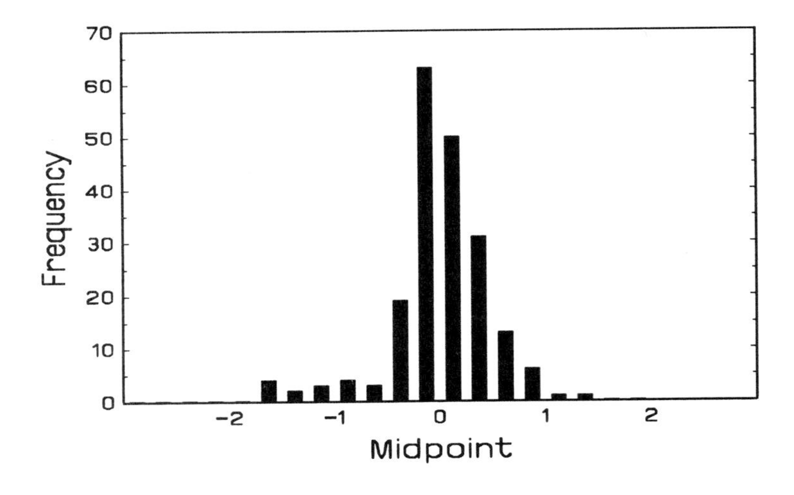

Figure 1.3: Empirical distribution of OLS residuals.

One way to test the normality assumption of u_t, and hence p_t, is to construct a histogram of the ordinary least squares residuals. This is presented in Figure 1.2 and shows that there is no evidence to reject the hypothesis that the distribution of u_t is unimodal. A more formal test of normality is given by the Jarque-Bera test. The computed value of the test statistic is 114.071. Comparing this value with a χ_2^2 critical value shows that the null hypothesis that the disturbance term u_t is normally distributed is rejected at standard significance levels.

A standard approach to solve the problems of autocorrelation, heteroscedasticity and ARCH is to reestimate (1.1) by generalised least squares. This solution is based, implicitly at least, on the assumption that the underlying distribution of p_t is unimodal, and it has been seen that there is no sign of multimodality. In fact, as is highlighted by the nonlinear modelling framework developed in Section 1.2, the rejection of normality using the Jarque-Bera test potentially provides the most illuminating information of all of the diagnostics reported so far, concerning the presence of multimodality.

1.1.2 A Structural Form

The linear approach described above began, as do many empirical studies, by writing a linear regression model relating the relevant variables; it was not derived directly from an economic theory of price determination. The specification in (1.1) may, however, be regarded as the reduced form of a model which is based on a structural form consisting of linear demand and supply functions. The price changes are assumed to be generated by shifts over time in these structural equations. Thus, suppose the demand, q_t^d, and supply, q_t^s, of the good in question can be written respectively as:

$$q_t^d = a_t - bp_t \tag{1.3}$$

$$q_t^s = c_t + dp_t \tag{1.4}$$

The intercepts, a_t and c_t, are assumed to be linear functions of the 'fundamental' exogenous variables x_t and y_t respectively, while the slope coefficients, b and d, are assumed to be constant, reflecting the idea that the price changes are produced by shifts in the curves rather than changes in their slopes, although this assumption can be easily relaxed. In equilibrium, $q_t^d = q_t^s$, so the equilibrium price in period t is given by the root of the equation:

$$p_t - \frac{a_t - c_t}{d + b} = 0 \tag{1.5}$$

1.1.3 A Distributional Specification

One way to introduce stochastics into the structural model in (1.3) and (1.4) is to include additive shocks in the intercept terms in the supply and demand functions. This has the effect of forcing p_t to deviate from its equilibrium price h_t. Assuming that the shock, v_t, is distributed as $N(0, 1)$, equation (1.5) generalises to:

$$p_t = h_t + \sigma_u v_t \tag{1.6}$$

where h_t is given by:

$$h_t = \frac{a_t - c_t}{d + b} \tag{1.7}$$

and is the equilibrium price. A comparison between (1.1) and (1.6) shows that the equilibrium price, h_t, is also the conditional mean $\alpha_0 + \alpha_1 x_t + \alpha_2 y_t$. It also shows that the linear equation given in (1.1) can be interpreted as the reduced form of this model.

The conditional price distribution, $f(p_t)$, which follows immediately from (1.6) and by use of the additive property of the normal distribution, is given by:

$$f(p_t) = N\left(h_t, \sigma_u^2\right) \tag{1.8}$$

The observed prices are treated as being random drawings from a conditional normal distribution whose mean in period t is given by the root of equation (1.5), which is of course $h_t = (a_t - c_t) / (d + b)$.

The form of the density function for the normal distribution is given by:

$$f(p_t) = \frac{1}{\sqrt{2\pi\sigma_u^2}} \exp\left(-\frac{1}{2\sigma_u^2}(p_t - h_t)^2\right) \tag{1.9}$$

It is useful to consider an alternative way of specifying this density which helps to make the contrast more clearly between linear models and the class of nonlinear models developed below. First, noting that $1/\sqrt{2\pi\sigma^2}$ can be expressed as $\exp\left(-\frac{1}{2}\log\left(2\pi\sigma^2\right)\right)$, $f(p_t)$ can be rearranged in the form of a second-order polynomial in p_t given by:

$$f(p_t) = \exp\left(\theta_{1,t}p_t + \theta_{2,t}p_t^2/2 - \eta_t^*\right) \tag{1.10}$$

where the coefficients on p_t and $p_t^2/2$ are respectively given by:

$$\theta_{1,t} = \frac{h_t}{\sigma_u^2} \tag{1.11}$$

$$\theta_{2,t} = -\frac{1}{\sigma_u^2} \tag{1.12}$$

and η_t^* is:

$$\eta_t^* = \frac{1}{2}\left(\frac{h_t^2}{\sigma_u^2} + \ln(2\pi\sigma_u^2)\right) \tag{1.13}$$

This term varies with t because h_t contains x_t and y_t. Essentially its role is to ensure that $f(p_t)$ qualifies as a density function; that is, it is required to have $\int_{-\infty}^{+\infty} f(p_t)\,dp_t = 1$. For this reason η_t^* is known as the 'normalising constant'. The term $\theta_{2,t}$ in (1.10) involves only the variance σ_u^2,which in the present model is assumed to be constant over time. The first term $\theta_{1,t}$ is a function of the exogenous variables and the variance term. This approach to the linear model, involving rewriting the standard expression for the normal distribution as a second-order polynomial, may at first sight appear cumbersome, but it provides the clearest way to see how it can be extended.

1.2 A Nonlinear Approach

1.2.1 The Generalised Exponential Family

A nonlinear approach can be developed as a natural extension of the above linear model by the simple addition of terms in the expression for the price distribution in equation (1.10). These terms may be powers of prices or transformations of prices such as reciprocals or logarithms. This is all that is required to produce the generalised exponential family of distributions. In the case where only additional powers are used, the price distribution, $f(p_t)$, now takes the form:

$$f(p_t) = \exp\left[\theta_{1,t}p_t + \theta_{2,t}p_t^2/2 + \theta_{3,t}p_t^3/3 + ... + \theta_{k,t}p_t^k/k - \eta_t\right] \tag{1.14}$$

where η_t is a normalising constant defined as

$$\eta_t = \ln \int \exp\left[\theta_{1,t}p_t + \theta_{2,t}p_t^2/2 + \theta_{3,t}p_t^3/3 + ... + \theta_{k,t}p_t^k/k\right] dp_t. \tag{1.15}$$

This corresponds to η_t^* above for the normal distribution and ensures that $f(p_t)$ qualifies as a density function. There is thus a direct comparison between equations (1.14) and (1.10).

The term in brackets in (1.14) may contain any number of terms in the polynomial, and as noted above may also contain terms involving transformed values of p, such as its logarithm. If only terms in p_t and p_t^2 appear in (1.14), then the price distribution reduces to the normal distribution by comparison with (1.10). If only terms in $\log(p_t)$, where p_t is strictly positive, and p_t appear, the distribution follows the unimodal gamma distribution; whereas if only p_t appears in (1.14), then it is a simple exponential distribution. Other forms contained within the generalised exponential family include the generalised gamma and generalised Student t distributions; see Cobb *et al.* (1983), Lye and Martin (1993a) and Creedy and Martin (1994b).

The terms $\theta_{j,t}$ in equation (1.14) are in general assumed to be functions of the fundamental or exogenous variables, x_t and y_t. For example, they could be written as linear functions such as:

$$\theta_{j,t} = \alpha_{j,0} + \alpha_{j,1}x_t + \alpha_{j,2}y_t \tag{1.16}$$

for $j = 1, 2, ..., k$. This means that over time the θs can change. This has the far-reaching implication that the form of the conditional price distribution can change over time. This occurs not necessarily as a result of a changing conditional mean, as in the case of the linear model, or a changing conditional variance, as in the case where there is heteroscedasticity, but as a result of changes in higher-order conditional moments such as skewness and kurtosis. More interestingly, there is now also a possibility that the distribution can change over time from being unimodal to being multimodal. In practice, the precise relationship between the θs and the exogenous variables is not known. However, a significant advantage of the approach is that a form of (1.14) can be derived from an economic model, as described in the following section.

First, the remainder of this section examines the condition under which the distribution has more than one mode and then considers the performance of the nonlinear approach when faced with the hypothetical data used earlier.

1.2.2 Bimodality and the Generalised Exponential

It is important to be able to identify whether or not the generalised exponential distribution displays multimodality in any time period. This information is particularly valuable because multimodality may be associated with large discrete jumps in the price, where a movement takes place from one mode to another. Such a jump in the price need not be associated with large changes in the fundamental, or exogenous, variable of interest. This type of phenomenon is observed during stock market crashes where smooth changes in the economic fundamentals, as represented by dividends, are associated with relatively larger movements in the share price; see, for example, Barsky and De Long (1993), Genotte and Leland (1990), Lim *et al.* (1995), and chapter 9 below. This phenomenon is also typical of foreign exchange markets where large swings in exchange rates are associated with relative smaller movements in the economic fundamentals; see chapter 8.

For the purpose of identifying bimodality, it is convenient to transform the model. This alternative form exhibits the same qualitative properties as the original model. To show this, consider (1.14) with $k = 4$, whereby:

$$f(p_t) = \exp\left[\theta_{1,t}p_t + \theta_{2,t}p_t^2/2 + \theta_{3,t}p_t^3/3 + \theta_{4,t}p_t^4/4 - \eta_t\right] \tag{1.17}$$

This can be transformed to a distribution in the variable w_t by using:

$$w_t = p_t\left(-\theta_{4,t}\right)^{0.25} + \theta_{3,t}\left(-\theta_{4,t}\right)^{-0.75}/3 \tag{1.18}$$

After some tedious algebra it can be shown that:

$$f(w_t) = \exp[\tau_{0,t}w_t + \tau_{1,t}w_t^2/2 - w_t^4/4 - \eta_t^{**}] \tag{1.19}$$

where the terms $\tau_{0,t}$ and $\tau_{1,t}$, are expressed as functions of the θs, and η_t^{**} is the normalising constant. Notice that the coefficient on $w_t^3/3$ is zero, so that the cubed term is eliminated and the coefficient on $w_t^4/4$ is unity. This type of variate transformation is known as a 'diffeomorphism'; in general, in a polynomial of order k, the $(k-1)$th term can be eliminated. For further discussion of this type of transformation, see Gilmore (1981).

The importance of the transformation in (1.18) is that the transformed distribution in (1.19) exhibits the same characteristics as the untransformed distribution in (1.17); in particular, if (1.17) exhibits bimodality then so will (1.19). The advantage of eliminating the cubic term is that it is easier to examine the conditions for multimodality. It is possible to show, as in Cobb *et al.* (1983, p.128), that a necessary and sufficient condition for (1.19) to exhibit bimodality involves the term, δ_t, defined by:

$$\delta_t = \frac{\tau_{0,t}^2}{4} - \frac{\tau_{1,t}^3}{27} \tag{1.20}$$

This is known as Cardan's discriminant, and it needs to be negative for (1.19) to have two modes and one antimode. As $\tau_{0,t}$ and $\tau_{1,t}$ are functions of the exogenous variables x_t and y_t, which vary over time, then δ_t also varies over time. Hence δ_t can change sign, say from positive to negative over the sample period, which means that the conditional distribution changes from being unimodal to being multimodal. Inspection of (1.20) also shows that a necessary condition for bimodality is that $\tau_{1,t}$ needs to be positive; for a discussion in the context of multiple equilibria, see Creedy and Martin (1993, pp.343-345). Cardan's discriminant is used throughout the present book.

1.2.3 Estimation and Prediction

The generalised exponential family, such as the model described in equations (1.14) and (1.16) or the simplified transformed version in equation (1.19), can be estimated using maximum likelihood methods. For a time series consisting of observations on $t = 1, 2, ..., T$ periods, the log-likelihood is

$$L(\Psi) = \sum_{t=1}^{T} f(p_t; \Psi) \tag{1.21}$$

where Ψ represents the set of parameters that need to be estimated. The maximum likelihood point estimates as given by $\widehat{\Psi}$ are those values which maximize $L(\Psi)$. Standard errors can be computed in the usual way from the hessian of the log-likelihood when evaluated at the maximum likelihood estimates. Asymptotic t-statistics can be constructed which are asymptotically distributed as $N(0,1)$ under the usual regularity conditions.

The log-likelihood in (1.21) is maximised using standard iterative gradient algorithms. This contrasts with the linear model where the first-order conditions for maximum likelihood are solved explicitly for the coefficients. While the use of an iterative algorithm to compute maximum likelihood estimates is standard, what is not standard for the model developed in this chapter is that special attention has to be given to the normalising constant, η_t. This is because, as inspection of (1.15) shows, in general η_t is different for each time period, and numerical methods of integration have to be used. Further, as the normalising constant is a function of the parameters, the first-order conditions for maximum likelihood estimation also contain the derivatives of η_t with respect to the parameters in Ψ. However, as mentioned earlier, the aim of this chapter is to introduce the approach, so the more technical aspects of estimation are not discussed here; see Cobb *et al.* (1983), Lye and Martin (1993a), and in particular chapters 3 and 5 below.

It has been mentioned that a standard approach to prediction with the linear model is simply to take the conditional mean of the dependent variable, for given values of x_t and y_t. With a normal distribution, this conditional mean corresponds to the single mode. With the nonlinear model, however, it is necessary to devise an appropriate 'convention' for obtaining predictions because of the possible multimodality of the distribution. One approach, called the 'global' convention, is based on choosing that value corresponding to the highest mode; see chapters 8, 9 and 15.

As when using the linear model, the 'true' data-generating model cannot be known to the investigator, and it is necessary to specify a form for the distribution. If there is no *a priori* information concerning its form (for example, using the type of model discussed in the following section), one approach is to adopt a general specification whereby the θ terms in (1.17) are regarded as being functions of both x_t and y_t. An appropriate order for the polynomial in p_t must also be chosen.

The results of estimating one of many potential nonlinear model specifications are as follows, where absolute values of t-statistics are given in brackets:

$$f(p_t) = \exp\left[\widehat{\theta}_{1,t}p_t + \widehat{\theta}_{2,t}p_t^2/2 + \widehat{\theta}_{3,t}p_t^3/3 + \widehat{\theta}_{4,t}p_t^4/4 - \widehat{\eta}_t\right] \tag{1.22}$$

where

$$\begin{aligned}
\widehat{\theta}_{1,t} &= \underset{(3.333)}{3.437} \\
\widehat{\theta}_{2,t} &= -\underset{(2.160)}{5.663} - \underset{(0.803)}{0.174}\, y_t + \underset{(1.805)}{0.220}\, x_t \\
\widehat{\theta}_{3,t} &= \underset{(4.945)}{3.805} + \underset{(6.684)}{0.304}\, x_t - \underset{(1.073)}{0.043}\, y_t \\
\widehat{\theta}_{4,t} &= -\underset{(7.343)}{2.261}
\end{aligned}$$

$\widehat{\eta}_t$ is the normalising constant which is written with a ^ to signify that it is a function of the estimated parameters, and $\widehat{\sigma}_u = 0.328$; $\overline{R}^2 = 0.973$; $DW = 2.051$; $BP(2) = 24.844$; $ARCH(1) = 0.946$; $RESET(1) = 6.609$, with the residuals defined as the difference between p_t and the predictions based on the global mode of (1.22) at each t.[2] In this specification a fourth-order polynomial is chosen, and the terms $\theta_{1,t}$ and $\theta_{4,t}$ are treated as constant. As explained in the next section, this specification is motivated by a supply and demand model where the slopes of the respective curves are assumed to be constant. A comparison of the adjusted coefficient of determination shows some marginal improvement in goodness of fit with $\overline{R}^2$ increasing from 0.947 for the linear model to 0.973 for the nonlinear model.

Inspection of the t-statistics in (1.22) suggests that the coefficient on the variable y_t in the function $\theta_{3,t}$ is insignificant. However, the coefficients on both x_t and y_t in the function $\theta_{2,t}$ appear to be insignificant. Faced with these results, a sensible approach is to re-estimate the model deleting only the variable y_t in $\theta_{3,t}$. This gives the following results:

$$\begin{aligned}
\widehat{\theta}_{1,t} &= \underset{(3.192)}{3.374} \\
\widehat{\theta}_{2,t} &= -\underset{(1.861)}{4.013} - \underset{(3.812)}{0.367}\, y_t + \underset{(1.357)}{0.186}\, x_t \\
\widehat{\theta}_{3,t} &= \underset{(4.307)}{3.260} + \underset{(6.083)}{0.294}\, x_t \\
\widehat{\theta}_{4,t} &= -\underset{(6.746)}{2.203}
\end{aligned} \tag{1.23}$$

The coefficient on y_t in $\theta_{2,t}$ has become significant, while that on x_t clearly remains insignificant. Dropping this last term leads to the following estimated model, where as before absolute values of t-statistics are given in brackets:

$$\begin{aligned}
\widehat{\theta}_{1,t} &= \underset{(3.260)}{2.819} \\
\widehat{\theta}_{2,t} &= -\underset{(3.622)}{5.563} - \underset{(3.878)}{0.380}\, y_t
\end{aligned} \tag{1.24}$$

[2]The degrees of freedom used in computing $\overline{R}^2$ equals the total number of parameters estimated. For the estimated nonlinear model in equation (1.22), the number of estimated parameters is 8. The diagnostics reported for the estimated nonlinear model in (1.22) as well as subsequent estimated nonlinear models are the same diasgnostics reported for the linear model. For the present purposes, these diagnostics are used as portmanteau tests for identifying the presence of nonlinearity by identifying if there is any additional information in the residuals. In the case of testing for $ARCH$ where the conditional distribution is from the generalised exponential family, see Lim *et al.* (1996).

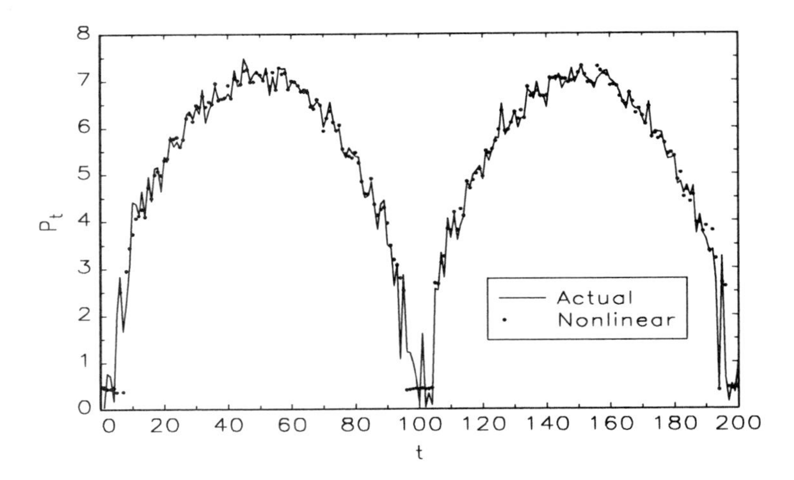

Figure 1.4: Actual and predicted p_t: nonlinear model in (1.24)

$$\begin{aligned} \widehat{\theta}_{3,t} &= \underset{(5.251)}{2.633} + \underset{(7.940)}{0.254}\, x_t \\ \widehat{\theta}_{4,t} &= -\underset{(8.117)}{1.982} \end{aligned}$$

with $\widehat{\sigma}_u = 0.379$; $\overline{R}^2 = 0.965$; $DW = 2.022$; $BP(2) = 26.572$; $ARCH(1) = 0.657$; $RESET(1) = 3.744$. The coefficients are now all significantly different from zero.

In terms of standard criteria, the adjusted coefficient of determination shows that there is a marginal improvement in goodness of fit over the linear model, while the DW statistic shows no evidence of autocorrelation. The $ARCH$ and $RESET$ test statistics show no evidence of nonlinearities, and while the BP statistic still points to some evidence of heteroscedasticity the degree of significance of heteroscedasticity has been dramatically decreased from the result reported for the linear model. Comparing the predictions of this model, shown in Figure 1.3, with the predictions of the linear model given in Figure 1.1, shows that the nonlinear model performs better in predicting the price in those periods where prices are relatively low.

More importantly, by using Cardan's discriminant as given in equation (1.20), the nonlinear model is able to identify periods of bimodality. This is shown by the negative values of Cardan's discriminant, as displayed in Figure 1.4. The bimodal properties of the model are further demonstrated in Figure 1.5, which gives snapshots of the distribution of prices conditional on specific values of x_t and y_t. Hence, it is concluded that the prices observed for the relevant periods shown in Figure 1.3 have been selected from an underlying distribution that is multimodal. This has been revealed by the non-linear approach, in contrast with the linear approach which, at best, suggests only that the tails of the homoscedastic price distribution are somewhat fatter than the normal distribution and that it is a little more peaked.[3] The distribution obtained by temporal aggregation may not, as in the case here, reveal the information that multimodality is present in some of the periods, particularly if there are relatively few periods of such multimodality.

It is shown in Section 1.4 that this is in fact the model used to generate the hypothetical data. The parameter estimates are close to those used to generate the data, and the three periods of multimodality identified in Figure 1.4, namely at the start, middle and end of the sample, are the correct periods. In particular, the number of points in time where the actual distribution is bimodal is 24, of which the estimated model predicts 17 correctly. The estimated model does not predict any periods of bimodality when the true distribution is unimodal. In comparison with the linear model, the non-linear model is able to capture a great deal more of the properties of the data generation process. The investigator does not need to know, *ex ante*, which periods display multimodality and involve jumps from one mode to another.

1.3 Data Generating Processes

The procedure followed in Section 1.2 is one in which the price distribution at any time is specified as a generalised exponential distribution which is conditional on the exogenous variables. Without knowing anything about the 'true' data generation process, it was shown how an investigator might select various specifications. The purpose of this section is to show how it is possible to start from an *a priori* specification of a model and generate the implied form of the distribution. Instead of the linear structural form of the

[3] In particular, Lye and Martin (1993b) show that a Lagrange multiplier test of the joint restrictions $\theta_3 = \theta_4 = 0$, in the density:

$$f(p_t) = \exp\left[\theta_1 p_t + \theta_2 p_t^2/2 + \theta_3 p_t^3/3 + \theta_4 p_t^4/4 - \eta\right]$$

is equivalent to the Jarque-Bera test of normality; see also chapters 3 and 4 for extensions. The implication of this result for empirical research is that when the normality assumption is rejected, this could be the result that the underlying distribution comes from the generalised exponential family which can exhibit multimodality.

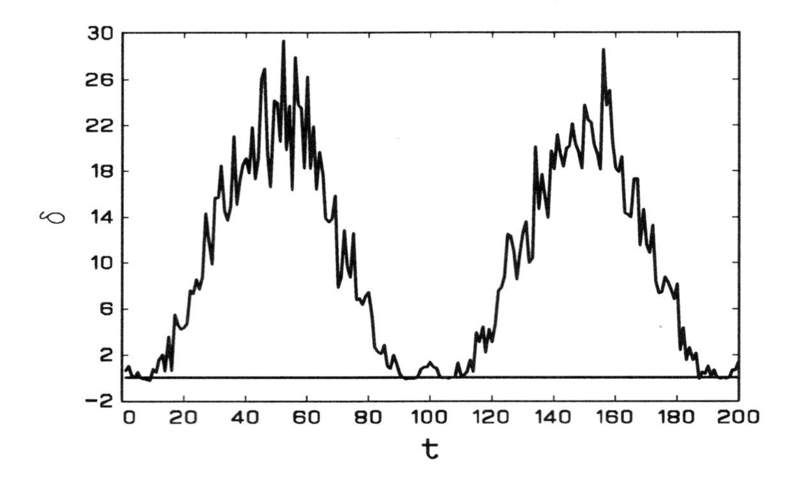

Figure 1.5: Cardan's discriminant: $\delta_t < 0$ implies bimodality

model where the equilibrium price is the unique root of the linear equation (1.5), it is possible to start with a nonlinear structure whereby the equation giving the equilibrium price may have more than one root. In addition, as shown in the following subsection, the stochastics can be introduced into the structure in a more complex manner.

1.3.1 The Stochastic Differential Equation

Suppose an economic model of price determination generates the equilibrium price as the root, or an appropriate root, of the equation:

$$\mu(p_t) = 0 \tag{1.25}$$

The function $\mu(p_t)$ may, for example, be a polynomial in p_t, where the coefficients on p_t are time-varying as they depend in some way on exogenous economic variables. This represents the deterministic form of the price model. For example, in the partial equilibrium supply and demand model discussed in Section 1.1 above, $\mu(p_t)$ is a linear function, $p_t - h_t$, with the equilibrium price given as a linear function of the exogenous variables. In that linear context, the observed price in each period is regarded as a random drawing from a normal distribution with the equilibrium price as the arithmetic

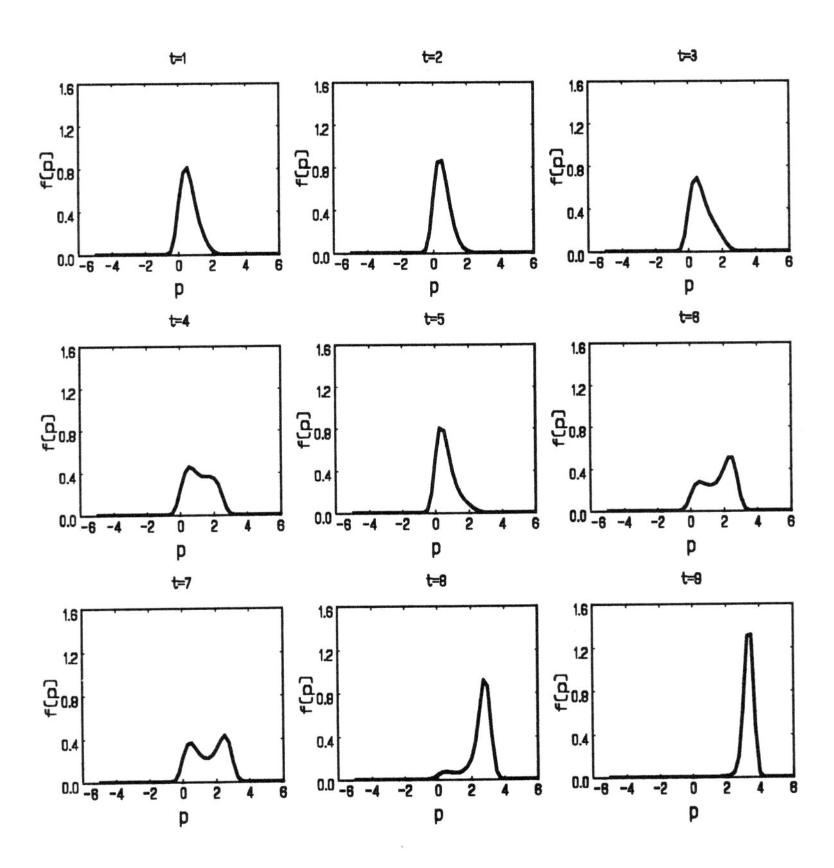

Figure 1.6: Temporal distribution snapshots: periods $t = 1, 2, ..., 9$

mean. Examples of alternative models which give rise to nonlinear forms of $\mu(p_t)$ are given below, but at this stage it is useful to concentrate on the stochastics of the general form of the model.

Stochastics can be introduced into the deterministic model in various ways. One approach is to assume that (1.25) represents the mean of a stochastic process and that deviations from the mean are regarded as arising from continuous additive stochastic shocks. The stochastic representation of the model is then given by:

$$\frac{dp_t}{dt} = -\mu(p_t) + \sigma_t v_t \tag{1.26}$$

where the stochastic term, v_t, is assumed to follow a normal distribution with zero mean and unit variance. This differential equation can be interpreted as an 'error correction model' whereby the price adjusts continuously from its mean as a result of short-run stochastic shocks, v_t. The negative sign means that on average the price adjusts over time back to the long-run equilibrium position, that is it 'error corrects'. Indeed, (1.25) can be regarded as a special case of (1.26) since in the absence of random shocks, $v_t = 0$, prices are in equilibrium and $dp_t/dt = 0$. The formulation in (1.26) may usefully be compared with the linear model as expressed in (1.6). The latter introduces stochastics simply by adding a normally distributed 'disturbance' term to the equilibrium price.

Since it is assumed that p_t and v_t are continuous random variables, it is convenient to express (1.26) in its stochastic continuous time form, which produces what is known as an Ito process (see Kamien and Schwartz, 1981, pp.243-244). This can be written as follows:

$$dp_t = -\mu(p_t)dt + \sigma(p_t)dW_t \tag{1.27}$$

In this formulation, $\mu(p_t)$ is the (instantaneous) mean given by (1.27) and $\sigma^2(p_t)$ is the (instantaneous) variance; the latter is in general also a function of p_t. The term dW_t is equal to $(v_t/\sigma(p_t))\, dt$, where W_t is known as a Wiener process, such that dW_t is distributed as $N(0, dt)$. This represents the continuous time analogue of the normal distribution. For a more formal discussion of the relationship between (1.26) and (1.27), see Brock and Malliaris (1982, pp.67-68).

1.3.2 Transitional and Stationary Distributions

For a given set of structural parameters, the imposition of the above stochastic process implies that p_t is a random variable with density function $f(p_t)$. In view of the fact that the structural parameters themselves may change as result of changes in any exogenous variables of which the structural parameters are functions, it is necessary to distinguish two types of distributions. First, if in the deterministic form of the model, adjustment of prices towards equilibrium is instantaneous, then the continuous stochastic process described in (1.27) results in a stable distribution. This can be referred to as a stationary distribution. Secondly, when adjustment is not instantaneous, the process can be regarded as moving through a sequence of transitional distributions before converging to the stationary distribution. It is important to stress that the stationary distribution is in general a function of exogenous variables. By changing the values of the exogenous variables this gives rise to a sequence of transitional distributions which converges to a stationary distribution. If the values of the exogenous variables do not change thereafter, realized observations on p_t will come from this stationary distribution.

The transitional distribution can be examined by considering a shock to the system resulting from an exogenous change in the structural parameters in $\mu(p_t)$. If adjustment is not instantaneous, then over time the process moves through a sequence of temporary equilibria. Associated with each temporary equilibrium is the transitional density of p_t. The dynamics of the process given by (1.27) are summarised by the Kolmogorov forward equation (see, for example, Cox and Miller, 1984, p.208):

$$\frac{\partial f_t(p_t)}{\partial t} = \frac{\partial}{\partial p_t}\left(\mu(p_t) f_t(p_t)\right) + \frac{1}{2}\frac{\partial^2}{\partial p_t^2}\left(\sigma^2(p_t) f_t(p_t)\right) \tag{1.28}$$

This is a partial differential equation which defines the transitional density, $f_t(p_t)$, at each point in time. Unfortunately, except for very simple expressions for both the mean and the variance, $\mu(p_t)$ and $\sigma^2(p_t)$, no analytical solution for this partial differential equation exists. The complexity of apparently simple processes is highlighted in Soong (1973). One procedure to overcome this problem is the concept of the stationary density mentioned above, and discussed by Soong (1973, pp.197-199).

The stationary density represents the stochastic analogue of long-run equilibrium in the deterministic model; see Brock and Malliaris (1982, pp. 106-108). It is possible to derive analytical expressions for the stationary density using general expressions for the mean and the variance. From (1.28), the stationary density, $f(p_t)$, is found by setting:

$$\frac{\partial f(p_t)}{\partial t} = 0 \tag{1.29}$$

This converts the partial differential equation in (1.28) to an ordinary differential equation in the stationary density $f(p_t)$ which is independent of time.

The general expression for the stationary density is derived as follows. First combine (1.28) and (1.29) to get:

$$0 = \frac{d}{dp_t}\left(\mu(p_t)f(p_t)\right) + \frac{1}{2}\frac{d^2}{dp_t^2}\left(\sigma^2(p_t)f(p_t)\right) \tag{1.30}$$

Expanding the second term using the product rule, (1.31) can be rewritten as:

$$0 = \frac{d}{dp_t}\left(\mu(p_t)f(p_t)\right) + \frac{1}{2}\frac{d}{dp_t}\left(f(p_t)\frac{d\sigma^2(p_t)}{dp_t} + \sigma^2(p_t)\frac{df(p_t)}{dp_t}\right) \tag{1.31}$$

Integrating both sides with respect to p_t gives a first-order linear differential equation:

$$k = \mu(p_t)f(p_t) + \frac{1}{2}\left(f(p_t)\frac{d\sigma^2(p_t)}{dp_t} + \sigma^2(p_t)\frac{df(p_t)}{dp_t}\right) \tag{1.32}$$

where k is a constant of integration. Solving for $df(p_t)/dp_t$ gives:

$$\frac{df(p_t)}{dp_t} = \frac{2k}{\sigma^2(p_t)} - \left(\frac{2\mu(p) + d\sigma^2(p_t)/dp_t}{\sigma^2(p_t)}\right) f(p_t) \tag{1.33}$$

Finally, the stationary density is given as the solution of this differential equation:

$$f(p_t) = \exp\left[-\int_0^{p_t}\left(\frac{2\mu(s) + d\sigma^2(s)/ds}{\sigma^2(s)}\right) ds - \eta_t\right] \tag{1.34}$$

where η_t denotes the appropriate normalising constant to ensure that $f(p_t)$ qualifies as a density. This density represents the generalized exponential distribution discussed in Section 1.2.

It is also possible to show that the number of modes and antimodes of the density is given by the number of roots of:

$$2\mu(p_t) + d\sigma^2(p_t)/dp_t = 0 \tag{1.35}$$

For the case where the variance is constant, $d\sigma^2(p_t)/dp_t = 0$, the modes and antimodes equal the stable and unstable equilibrium values respectively. In the general case where the variance is not constant, the modes and antimodes are displaced from the equilibrium points by the factor $d\sigma(p_t)/dp_t$. The variance of the process, $\sigma^2(p_t)$, plays an important part in determining the form of the stationary density; see Cobb *et al.* (1983).

The result shown in equation (1.34) is extremely useful. It can be applied directly to a wide variety of contexts, where the forms of $\mu(p_t)$ and $\sigma^2(p_t)$ are given from basic economic arguments. The way in which it can be applied is perhaps best illustrated by considering a number of examples, and several special cases are therefore discussed in the following section.

1.4 Examples of Nonlinear Models

1.4.1 Partial Equilibrium Analysis

In order to illustrate how the above results can be used, consider again the special case where the equilibrium price in a partial equilibrium model is given by the root of equation (1.5). This can be written as $\mu(p_t) = p_t - h_t$, where the variable h_t is understood to be a function of the exogenous variables. Suppose also that the variance is constant, so that $\sigma^2(p_t) = \gamma$. Substituting these assumptions into (1.34) gives the stationary distribution:

$$f(p_t) = \exp\left[-\int_0^{p_t}\left(\frac{2(s-h_t)}{\gamma}\right)ds - \eta_t\right] \tag{1.36}$$

Integrating and collecting terms gives the result that:

$$f(p_t) = \exp\left[\theta_{1,t}p_t + \theta_{2,t}p_t^2 - \eta_t\right] \tag{1.37}$$

with:

$$\theta_{1,t} = 2h_t/\gamma \tag{1.38}$$
$$\theta_{2,t} = -1/\gamma \tag{1.39}$$

Comparison with (1.10) shows that this stationary distribution corresponds to the familiar normal distribution. Substitution of these assumptions into (1.35) shows, not surprisingly, that this distribution has just one mode. This example serves to illustrate the link between the linear and nonlinear models: the former is just a special case of the latter.

1.4.2 Variance Proportional to Price

A modification of the model of the previous subsection is obtained by allowing the variance, $\sigma^2(p_t)$, to be proportional to the price instead of being constant, so that $\sigma^2(p_t) = \gamma p_t$. Substitution into (1.34) gives:

$$f(p_t) = \exp\left[-\int_0^{p_t}\left(\frac{2(s-h_t)+\gamma}{\gamma s}\right)ds - \eta_t\right] \tag{1.40}$$

The integration of the term in $1/s$ introduces $\log(p_t)$, so that the stationary distribution becomes:

$$f(p_t) = \exp\left[\theta_{1,t}\ln(p_t) + \theta_{2,t}p_t - \eta_t\right] \tag{1.41}$$

with:

$$\theta_{1,t} = (1 - 2h_t/\gamma) \tag{1.42}$$

$$\theta_{2,t} = -2/\gamma \tag{1.43}$$

This is the unimodal gamma distribution. The term $\theta_{1,t}$ is a function of the exogenous variables which produce the shifts in the structural equations, while the term $\theta_{2,t}$ depends only on the variance term γ.

1.4.3 A Structural Model with Multiple Equilibria

The previous two examples have involved the linear reduced form of a partial equilibrium structural model. Consider next an exchange model involving two goods, where the demands for these goods at time t, $q_{i,t}^d$ $(i = 1, 2)$ are assumed to be linear functions of the relative price, where now $p_t = p_{1,t}/p_{2,t}$, such that:

$$q_{1,t}^d = a_t - b_t p_t \tag{1.44}$$

$$q_{2,t}^d = c_t - d_t p_t^{-1} \tag{1.45}$$

In this formulation, the intercepts and slopes of the two demand curves, $a_t, ..., d_t$, are given t subscripts to indicate that they can change over time. Each parameter may be specified as a function of a set of exogenous variables. The fundamental reciprocal demand and supply concept means that only the two demand curves need to be specified. This is in fact a special case of an exchange model which has a long pedigree; see Creedy (1992, 1996) and Creedy and Martin (1994b, c). This type of framework is used in chapters 2 and 6 below. If $q_{2,t}^s$ denotes the supply of good 2 in period t, then

$$p_{1,t}/p_{2,t} = q_{2,t}^s/q_{1,t}^d \tag{1.46}$$

so that the supply of good 2 in period t is given by $p_t q_{1,t}^d$. Equating this with the demand for good 2, $q_{2,t}^d$, gives the equilibrium price as the root or roots of:

$$b_t p_t^3 - a_t p_t^2 + c_t p_t - d_t = 0 \tag{1.47}$$

The reduced form of the model in (1.47) is thus a cubic, which can have up to three distinct roots. Substitution into (1.34), with the additional assumption that the variance, $\sigma^2(p_t)$, is constant at γ, gives the result that the equilibrium distribution of the relative price is:

$$f(p_t) = \exp\left[-\frac{2}{\gamma}\int_0^{p_t}(b_t s^3 - a_t s^2 + c_t s - d_t)ds - \eta_t\right] \tag{1.48}$$

The integration of this expression gives:

$$f(p_t) = \exp\left[\theta_{1,t}p_t + \theta_{2,t}p_t^2/2 + \theta_{3,t}p_t^3/3 + \theta_{4,t}p_t^4/4 - \eta_t\right] \tag{1.49}$$

where:

$$\theta_{1,t} = 2d_t/\gamma \tag{1.50}$$
$$\theta_{2,t} = -2c_t/\gamma \tag{1.51}$$
$$\theta_{3,t:} = 2a_t/\gamma \tag{1.52}$$
$$\theta_{4,t} = -2b_t/\gamma \tag{1.53}$$

In this model, shifts in the two demand curves are produced solely by variations over time in the intercepts, a_t and c_t, resulting from changes in the exogenous variables which influence them. Hence the two terms $\theta_{2,t}$ and $\theta_{3,t}$, can also be expressed as the same types of function of the exogenous variables. If the slope coefficients, b_t and d_t, are constant over time, so that $b_t = b$ and $d_t = d$ for all t, then inspection of (1.50) and (1.53) shows that the terms $\theta_{1,t}$ and $\theta_{4,t}$ are also constant over time. This is indeed the basic model used to generate the hypothetical data used above, as explained in the next section.

It can also be seen by comparison with the previous example that the alternative assumption that the variance term is proportional to the price, so that $\sigma^2(p_t) = \gamma p_t$, leads to what is referred to as a generalised gamma distribution because of the introduction of the term in $\log(p_t)$; see Cobb *et al.* (1983). The number of modes and antimodes of the distribution of relative prices given in (1.49) is given by the number of roots of (1.35). It can be shown that the antimode corresponds to the root of (1.47) which is unstable: see Creedy and Martin (1993).

The fundamental result in (1.34) can therefore be applied to a large variety of economic models. So far, the discussion has been in terms of time series models, but the approach can also be used in cross-sectional contexts, as shown in Part II. If the economic model does not give rise to a convenient polynomial for $\mu(p_t)$, it is usually possible to convert it to a polynomial of appropriate order using an appropriate Taylor series expansion; see Creedy and Martin (1993, pp.346-347).

1.5 Generating the Hypothetical Time Series

This section describes the model and method used to generate the hypothetical times series used in earlier section.

1.5.1 The Model

The basic model used is the exchange model introduced in Section 1.4, involving two goods in a general equilibrium context. The feature of the model is that the demands are linear functions of the relative price, p_t, and the reduced form is the cubic equation in (1.47). It is assumed that the intercepts, a_t and c_t, of the demand curves in equations (1.44) and (1.45) are functions respectively of the exogenous variables x_t and y_t, such that:

$$a(x_t) = 3 + 0.25x_t \tag{1.54}$$

$$c(y_t) = 5 + 0.5y_t \tag{1.55}$$

The slope terms in equations (1.44) and (1.45) are, however, assumed to be constant at:

$$b = 2 \tag{1.56}$$

$$d = 3 \tag{1.57}$$

It is also assumed that the variance $\sigma^2(p_t)$ in (1.48), is constant where

$$\sigma^2(p_t) = \gamma = 2 \tag{1.58}$$

It is also necessary to specify the processes used to generate hypothetical values of the exogenous variables x_t and y_t. The series x_t is generated as:

$$x_t = 15 + 1.4t - 0.014t^2 + u_{x,t} \tag{1.59}$$

for $1 \leq t \leq 100$. For $100 < t \leq 200$, t and t^2 on the right-hand side of (1.59) are simply replaced by $t - 100$ and $(t - 100)^2$. The error term $u_{x,t}$ is distributed as $N(0, 1)$. The series y_t is generated as:

$$y_t = 10 + u_{y,t} \tag{1.60}$$

where $u_{y,t}$ is $N(0, 1)$.

The choice of parameters and specifications of the exogenous variables in the simulation experiment give rise to 24 points in time where the price distribution is bimodal. This results in an unconditional distribution of p_t

through temporal aggregation of the data which is unimodal. It has been seen that such an approach can give a misleading indication of the form of the appropriate conditional distributions. The simulation experiments could also be run where the number of points of time for which the conditional distribution is bimodal is increased. This, however, would raise the possibility that the unconditional distribution is also bimodal, which is not typical of actual data.

1.5.2 Simulation procedure

The above model can be used to provide a simulated price series using an approach known as the 'the inverse cumulative density method'. This involves constructing the cumulative density and using a uniform random number generator. The approach consists of the following steps. First, using the expressions for (1.54) to (1.58) in (1.50) to (1.53), the density function in (1.49) for the random variable p_t is:

$$f(p_t) = \exp\left[3p_t - (5 + 0.5y_t)p_t^2/2 + (3 + 0.25x_t)\,p_t^3/3 - 2p_t^4/4 - \eta_t\right] \quad (1.61)$$

where as before the normalising constant for each period, η_t, is chosen to ensure that the condition $\int f(p_t)dp_t = 1$, is satisfied. Secondly, a cumulative density is constructed:

$$F(u) = \int_{-\infty}^{u} f(p_t)dp_t \quad (1.62)$$

As there is no explicit expression for the cumulative density, it is computed numerically by replacing the integral sign with a summation sign and choosing small steps for dp_t. Thirdly, using the property that the cumulative density is monotonic and starts at zero and ends at unity, a uniform random number generator can be used to generate random numbers for u in (1.62) on the interval $[0,1]$. Associated with each random number, u, in (1.62) is a value for p, which acts as the random price used.

To obtain a time series of p_t for the periods $t = 1, 2, ..., 200$, the approach consists of initially setting $t = 1$ in (1.61). This amounts to substituting in the realised values of x_1 and y_1 in (1.61). By using (1.62) with $t = 1$, the realised price p_1, is generated. The time step is incremented to $t = 2$, and the process repeated until the time last time step, $t = 200$. Hence at each t the density used to generate the realisation of p_t varies as a result of the variation in the values of x_t and y_t when substituted in (1.61).

1.6 Conclusions

This chapter has introduced the approach to nonlinear modelling based on the use of the generalised exponential family of distributions. The considerable flexibility of the approach was illustrated using hypothetical data based on an economic model which exhibits multiple equilibria for certain periods of time and a unique equilibrium for other periods. It was shown that the distributional analogue of multiple equilibria is multimodality.

An important advantage of this framework is that discrete jumps can be modelled in a natural way without the need to identify the timing of jumps *ex post.* This is particularly valuable for policy analysis. For example, one of the fundamental exogenous variables of the model may be a policy variable that can be controlled by the government. It is possible that a very small change in such a variable can in some time periods lead to a very large jump in the dependent variable, depending on the values of other exogenous variables. The same type of change in the policy variable may at other times have only a small effect on the dependent variable. Such jumps can be predicted even if they have never been observed during the sample period.

The framework developed in this chapter also has the advantage of explaining how smooth changes in market fundamentals can give rise to large and sudden changes in prices, as is characteristic of foreign exchange and stock markets. This result is important as it suggests that when standard economic models break down, during stock market crashes for example, this need not be the result of a misspecification of the variables used to model the market fundamentals, but it may represent the adoption of an incorrect functional form.

The introduction of economic assumptions into nonlinear models was explained, and it was shown, with the addition of stochastics, how an explicit form for the distribution of the dependent variable can be derived. The approach has considerable potential in a wide variety of economic contexts, as subsequent chapters in this book demonstrate.

Part II

Cross-sectional Applications

Chapter 2

A Model of Income Distribution

John Creedy, Jenny N. Lye and Vance L. Martin

There is a substantial literature seeking to explain the form of the distribution of earnings in terms of the outcome of some kind of random process. Approaches involve the application of a form of the central limit theorem to the sum of random variables, or the analysis of the results of applying a transition matrix to an arbitrary initial distribution, in terms of the characteristic vector corresponding to the unit root of the matrix. In each case the 'final' distribution is independent of the initial distribution. These include the famous 'law of proportionate effect' of Gibrat, discussed by Aitchison and Brown (1957) and Brown (1976) as a possible genesis of the lognormal form, and the alternative model of Champernowne (1953) which produces the Pareto distribution; see also Mandelbrot (1960), Klein (1962), Hart (1973), and Shorrocks (1975). For a useful non-technical discussion, see Phelps Brown (1977, pp.290-294).

These models have, perhaps not surprisingly, been criticised on the grounds that they are 'purely statistical' and thereby have a negligible degree of economic content; see in particular Mincer (1970) and Lydall (1979, pp.233-236). Lydall (1976, p.19) has argued, for example, that 'this type of stochastic theory is . . . not scientific in the usual sense. The "explanation" which it offers is at a very superficial level and does not explain any of the real factors - economic or other - which are responsible for the shape of the distribution.'

This chapter explores an approach to modelling the earnings distribution which involves both stochastic and economic components. Individuals' earnings are seen as resulting from a simple market supply and demand model in which individual earnings are deterministic, though the model can have multiple solutions. A stochastic error correction mechanism is then imposed and the resulting form of the distribution of earnings is derived, where the distributional analogue of multiple equilibria in the deterministic form is multimodality. In particular the model generates a flexible functional form described as the generalised gamma distribution. Special cases of this general distribution include the gamma, exponential and Weibull distributions as well as the power distribution. An advantage of the approach is that changes in the form of the distribution over time can be modelled explicitly in terms of changes in commodity and labour market conditions since the parameters governing the characteristics of the distribution are shown to be functions of the underlying economic parameters of the model. This makes it possible to give a clearer economic interpretation of the distribution of personal income and the reasons for changes in inequality over time.

The basic model is introduced in Section 2.1, which describes the labour demand and supply components and the stochastic specification. In view of the exploratory nature of the analysis, the model is very simple; however, it is rich enough to generate a distribution of personal earnings from first principles which is sufficiently flexible to capture a range of observed distributional characteristics. Section 2.2 describes the method of estimation, where both least squares and maximum likelihood methods are discussed. The generalised gamma distribution is applied in Section 2.3 to estimating the distribution of alternative earnings models.

2.1 The Model

2.1.1 Labour Demand and Supply

Consider a market in which all individuals are assumed to be in the labour force. Individual i's real earnings are denoted y_i, which are by definition the product of the wage rate per hour which the individual can obtain, w_i, and the number of hours worked, n_i. Thus:

$$y_i = w_i n_i \tag{2.1}$$

The individual's labour supply may of course be derived using a standard utility maximising approach in which the arguments of the utility function are consumption and leisure. For present purposes it is convenient to adopt the approach discussed by Robbins (1930) which exploited the reciprocal nature of labour supply in terms of the demand for goods, or real income. Suppose that the individual's demand for goods can be expressed as the following simple function of the real wage, the reciprocal of which represents the price of goods in terms of labour:

$$x_i^d = \alpha_i - \beta_i w_i^{-1} \tag{2.2}$$

The associated labour supply function is therefore given by:

$$n_i^s = x_i^d / w_i \tag{2.3}$$

The labour demand function facing the individual can be viewed as resulting from profit maximising, such that the optimal solution involves the demand for labour being expressed as a function of the real wage. For a linear labour demand schedule, the relationship is represented as:

$$n_i^d = \delta_i - \gamma_i w_i \tag{2.4}$$

It is required to solve these equations for the individual's real income, y_i. This can be done as follows. Use the demand function in (2.2) to express w_i in terms of x_i^d, so that $w_i = \beta_i/(\alpha_i - x_i^d)$, and substitute the result into (2.3) to give:

$$n_i^s = x_i^d(\alpha_i - x_i^d)/\beta_i \tag{2.5}$$

This is in fact the 'offer curve' of the individual, since it expresses the supply of labour directly in terms of the demand for goods: this has the typical 'backward bending' shape of standard offer curves. In equilibrium, the supply of labour given by (2.5) must equal the demand as in (2.4), where in the latter, w_i is also expressed in terms of x_i^d. This equilibrium condition gives, after some rearrangement, the following cubic, writing $x_i = x_i^d$:

$$x_i^3 - 2\alpha_i x_i^2 + (\alpha_i^2 + \beta_i\delta_i)x_i - (\beta_i\delta_i\alpha_i - \gamma_i\beta_i^2) = 0 \tag{2.6}$$

In this model real consumption is equivalent to real earnings since there is no role for savings, so it is only necessary to subsitute for $y_i = x_i$ in (2.6) to obtain the equation for earnings. Hence:

$$y_i^3 - 2\alpha_i y_i^2 + (\alpha_i^2 + \beta_i\delta_i)y_i - (\beta_i\delta_i\alpha_i - \gamma_i\beta_i^2) = 0 \tag{2.7}$$

This summarises the equilibrium properties for the ith individual in the market. There may be a unique equilibrium when (2.7) has a single real root, or multiple equilibria when it exhibits two or three real roots. In cases where there are three real roots, two of the equilibria can be shown to be stable; and if there are two real roots, only one is stable.

It can also be shown that the equilibrium wage rate corresponds to the root or roots of a cubic equation: subsitute for x_i^d in (2.3) using (2.2), and equate the result to the labour demand given by (2.4). The equilibrium properties of the determination of a price, in this case the price of labour per hour, rather than the product of price and quantity, in this case real earnings, have been examined in detail in Creedy and Martin (1993, 1994b, c).

The above supply and demand model is obviously very simple, but for present purposes it is sufficient for demonstrating how such an economic model can be used as the basic component of a distributional model. Further complications may provide a more detailed foundation for the functions used, and may even change the order of the polynomial in (2.7). A further possibility is that the equilibrium real wage may not be expressed as the root (or roots) of a polynomial, but it would be possible to rewrite the condition as a polynomial using a Taylor series expansion. The major elements of the approach followed below would thus remain unchanged, and examples of alternative specifications are discussed in Section 1.2.

2.1.2 The Stochastic Specification

The analysis so far is deterministic and describes the equilibrium properties of the level of earnings of an individual, with a specified demand function for goods (real earnings) and facing a demand function for labour. If some joint distribution of the parameters $\alpha_i, \beta_i, \delta_i, \gamma_i$ were specified, it would in principle be possible to examine the implied distribution of real earnings, though this would be extremely complex and some method of handling the multiple equilibria would be required. It is most unlikely that any analytical expression for the distribution could be obtained following such an approach. Furthermore, it would in some sense simply be 'pushing back' the explanation of the form of the distribution one stage, in that some rationale for the joint distribution of the parameters among individuals would be required.

The present approach is instead based on the view that the parameters of each individual's demand and supply functions are subject to random or stochastic 'shocks' which lead to movements of earnings away from their equilibrium values. However, earnings are subsequently assumed to return or 'error correct' to their values which held before the shocks. Given the possibility of multiple equilibria, even small shocks can lead to large jumps in the value of real earnings over time. A standard representation of such a stochastic process involves the following expression, written in terms of the proportionate change in the individual's real earnings:

$$\frac{dy_i}{y_i} = -\left(y_i^3 - 2\alpha_i y_i^2 + (\alpha_i^2 + \beta_i\delta_i)y_i - (\beta_i\delta_i\alpha_i - \gamma_i\beta_i^2)\right) dt + \sigma_i dZ_i \quad (2.8)$$

Here Z_i is a Wiener process with the property that dZ_i is normally distributed as $N(0, dt)$ and σ_i is the standard deviation of dy_i/y_i which for simplicity is assumed to be constant; however, more elaborate specifications can be accommodated. For further details of this type of process, see Cox and Miller (1984).

The formulation in (2.8) applies to each individual in the market. Its continued application, following the approach described below, would generate the form of the probability density of each individual's earnings resulting from changes over time in response to the stochastic shocks. The informational requirements of such an approach are of course enormous, and ultimately it is required to generate the distribution of real earnings defined over all individuals. Thus to continue to allow for such a large degree of heterogeneity would not be helpful.

This Gordian knot may, however, be cut by introducing an assumption of population homogeneity. Thus the random shocks applying to individuals are assumed to be drawn from essentially the same distribution: this assumption is of course precisely the same as that used by all the stochastic process models mentioned above. Furthermore, the individuals are homogeneous to the extent that they have the same demand and supply functions. However, it will be shown below that some heterogeneity can be introduced by allowing some of the structural parameters to be functions of various characteristics such as labour market experience and education. At any moment in time, individuals are affected by quite different shocks; they are simply assumed to originate from the same distribution. This simplification makes it possible to drop the individual subscripts from equation (2.8) and to regard the resulting distribution as applying over all individuals in the market. The form of the distribution will be seen to depend on the precise nature of the demand and supply schedules and the characteristics of the Wiener process. The only distributional assumption made in the model is that of the normality of dZ.

If the density function of real income is written as $f(y)$, then (2.8) gives rise to the following Kolgomorov forward equation (see Cox and Miller, 1984):

$$\frac{\partial f}{\partial t} = \frac{\partial}{\partial y}\left(y^4 - 2\alpha y^3 + (\alpha^2 + \beta\delta)y^2 - (\beta\delta\alpha - \gamma\beta^2)y\right) + \frac{1}{2}\frac{\partial^2}{\partial y^2}\left(\sigma^2 y^2\right) \quad (2.9)$$

This is a partial differential equation which has the solution:

$$f(y) = \exp\left[h(y,t;\theta) - \eta(t;\theta)\right], 0 < y < \infty \quad (2.10)$$

where θ is a vector of parameters which, in turn, are functions of the structural parameters $\alpha, \beta, \delta, \gamma$, and $\eta(t;\theta)$ is the normalising constant given by:

$$\eta(t;\theta) = \ln \int \exp\left[h(y,t;\theta)\right] dy \quad (2.11)$$

The function $h(y,t;\theta)$ is a general expression which unfortunately has no closed form solution. This means that it is not possible to derive an analytical solution of the transitional density $f(y)$ except for some very special cases of the model given above; for a general discussion of the problems, see Soong (1973). This form of $f(y)$ in (2.10) is known as a transitional distribution since it describes the evolution over time of the distribution of earnings from an initial equilibrium distribution after a change in the parameters of the demand and supply functions caused by some shock.

This process will converge to the 'stationary' distribution, which is obtained by setting $\partial f/\partial t = 0$ in (2.9). This is the approach followed below. Thus, it must be acknowledged that the same type of issue arises here as with all the other types of stochastic process models, namely that of the speed of adjustment. Critics of the stochastic models have argued that any observed distribution is unlikely to be an equilibrium distribution, so that the results of such processes cannot strictly be applied. In addition, Shorrocks (1975) has used numerical simulation methods, finding that the 'half life' of such processes is typically many 'periods'.

A substantial advantage of the stationary distribution is that, unlike the transitional distribution, the stationary distribution can in general be derived analytically. By setting $\partial f/\partial t = 0$ in (2.9), this converts the partial differential equation into an ordinary first order differential equation which can be solved by standard integration methods to yield a closed form solution. For a general solution of the stationary distribution for general specifications corresponding to (2.8), see chapter 1. For the stochastic differential equation in (2.8), the stationary density, $f^*(y)$, is:

$$f^*(y) = \exp\left[-\Phi(y) - \eta^*\right] \tag{2.12}$$

where

$$\Phi(y) = \frac{2y^3/3 - 2\alpha y^2 + 2(\alpha^2 + \beta\delta)y - 2(\beta\delta\alpha - \gamma\beta^2 + 2\sigma^2)\ln(y)}{\sigma^2}$$

and

$$\eta^* = \ln \int_0^\infty \exp\left[-\Phi(s)\right] ds$$

is the normalising constant. The integration range is defined over the interval $(0, \infty)$ to ensure that the variance of the stochastic differential equation is non-negative.

The density given by (2.12) is now written more compactly as:

$$f^*(y) = \exp\left[\theta_1 \ln(y) + \theta_2 y + \theta_3 y^2 + \theta_4 y^3 - \eta^*\right], 0 < y < \infty \tag{2.13}$$

where the θs are functions of the structural form parameters, obtained by comparing coefficients in (2.12) and (2.13), and:

$$\eta^* = \ln \int_0^\infty \exp\left[\theta_1 \ln(s) + \theta_2 s + \theta_3 s^2 + \theta_4 s^3\right] ds \qquad (2.14)$$

The distribution given by (2.13) is the generalised gamma. Creedy and Martin (1994) have shown that the same form applies in the context of price distribution models, using a simple general equilibrium model similar to that used here. It can be unimodal which corresponds to the situation where the economic model given by (2.1) to (2.4) contains a unique equilibrium. Unlike the standard gamma distribution proposed by Salem and Mount (1974) to describe income distributions, this generalised gamma can also be multimodal. This situation arises when the model has multiple equilibria whereby the modes of the density correspond to the stable equilibria and the antimode corresponds to the unstable equilibrium. The flexibility of the generalised gamma distribution to model alternative empirical earnings distributions arises from the inclusion of the higher order terms, which enable such properties as skewness and kurtosis to be modelled adequately. Marron and Schmitz (1992) give an example of a distribution displaying multimodality.

The above approach can be applied to a single cross-section of earnings, or to a time series of cross-sections. In the latter case the parameters may be regarded as functions of exogenous variables which change over time. An advantage of this framework is that it provides a method of modelling the influence of macroeconomic variables on the personal distribution of earnings. This also introduces the interesting possibility that the form of the distribution of earnings could change in distinct ways, while the underlying generating process is unchanged. This represents a potentially fruitful area of future research.

A comparison of (2.12) and (2.13) shows that the five structural parameters $(\sigma, \alpha, \beta, \gamma, \delta)$ cannot all be recovered from the θ_is. The emphasis in the present paper is on the form of the distribution and hence the distributional parameters θ, so the lack of identifiability of the structural parameters is not considered to be a problem. However, as shown below, it is possible to discriminate between alternative classes of structural models through tests on the form of the distribution.

2.1.3 Alternative Specifications

The theoretical model discussed in Section 2.1 involves linear demand and supply functions. However, alternative specifications can be explored. For example, suppose that the goods demand function is rewritten as:

$$w_i = \alpha_i + \beta_i \ln(x_i^d) \tag{2.15}$$

Notice that when using a nonlinear specification of equation (2.2), it is simpler to write w as a function of x^d. This provides a tractable formulation in view of the need to solve the model for y. Combining (2.15) with (2.1), (2.3) and (2.4) gives real earnings as the root or roots of:

$$y_i + \gamma_i \beta_i^2 (\ln(y_i))^2 + \beta_i(2\gamma_i\alpha_i - \delta_i)\ln(y_i) + (\gamma_i\alpha_i - \delta_i)\alpha_i = 0 \tag{2.16}$$

which corresponds to (2.7). This gives rise to a stationary distribution of the form:

$$f^*(y) = \exp\left[\theta_1 (\ln(y))^3 + \theta_2 (\ln(y))^2 + \theta_3 \ln(y) + \theta_4 y - \eta^*\right], 0 < y < \infty \tag{2.17}$$

where the θ_is are functions of the structural parameters. This is a generalised lognormal distribution.

Another example consists of specifying the labour demand function in reciprocal form as:

$$n_i^d = \delta + \frac{\gamma}{w_i} \tag{2.18}$$

When combined with equations (2.1) to (2.3), this gives rise to a quadratic of the form:

$$y_i^2 - y_i(\gamma + \alpha) + (\delta\beta + \gamma\alpha) = 0 \tag{2.19}$$

Alternatively, if (2.2) is replaced by:

$$x_i^d = \alpha + \beta w_i \tag{2.20}$$

and is combined with equation (2.4), the following quadratic results:

$$\gamma y_i^2 - y_i(\delta\beta - \beta^2 + 2\alpha\gamma) + \alpha(\delta\beta + \gamma\alpha) = 0 \tag{2.21}$$

Finally, the combination of equation (2.18) and (2.20) leads to the linear form:

$$y_i(\beta - \delta) + (\alpha\delta - \beta\gamma) = 0 \tag{2.22}$$

An interesting feature of this equation is that the resulting distribution is the standard gamma distribution which is a special case of (2.13) with $\theta_3 = \theta_4 = 0$. The generalised gamma distribution also contains as special cases a number of other distributions which have been used in the empirical modelling of earnings distributions: the exponential distribution ($\theta_1 = \theta_3 = \theta_4 = 0$), the Weibull distribution ($\theta_2 = \theta_4 = 0$), and the power distribution ($\theta_2 = \theta_3 = \theta_4 = 0$). Where each distribution has been used in the past, a special estimation method has been devised, as in Salem and Mount (1974). The present approach thus has the substantial advantage of nesting several distributions which can therefore be consistently estimated and compared. In the case of the generalised lognormal distribution, as with the generalised gamma distribution, it is possible to devise theoretical models which give rise to special cases. For example, the standard lognormal distribution arises when $\theta_1 = \theta_4 = 0$, and the gamma distribution when $\theta_1 = \theta_2 = 0$.

2.2 Estimation Methods

This section presents procedures for estimating the theoretical income distributions derived above. Two cases need to be distinguished depending upon whether the structural form parameters $\alpha, \beta, \delta, \gamma$ are constant or are functions of individual characteristics. In the former case, the derived earnings distribution is referred to as an unconditional distribution, whilst for the latter, a set of conditional distributions are estimated. These cases are discussed in turn below. The generalised gamma distribution given by (2.13) represents a special case of the generalised exponential family. Further discussion of estimation is given in chapter 5, and for those cases where the data are truncated, the methods discussed in chapter 3 can be used.

2.2.1 The Unconditional Distribution

From the discussion above, a general form of the density for y is:

$$f^*(y) = \exp\left[\sum_{i=1}^{M} \theta_i \psi_i(y) - \eta\right] \tag{2.23}$$

where $\psi_i(.)$ is some general function depending on the density type, M is the number of terms in the density depending on the specification of the model, and η is the normalising constant which is given by:

$$\eta = \ln \int \exp\left[\sum_{i=1}^{M} \theta_i \psi_i(y)\right] dy \tag{2.24}$$

The $\psi_i(.)$ in the case of the generalised gamma income distribution are $\psi_1(y) = \ln(y)$, $\psi_2(y) = y$, $\psi_3(y) = y^3$, and $\psi_4(y) = y^3$, by comparison with (2.13). In the case of the generalised lognormal distribution, the $\psi_i(.)$

are $\psi_1(y) = (\ln(y))^3$, $\psi_2(y) = (\ln(y))^2$, $\psi_3(y) = \ln(y)$, and $\psi_4(y) = y$, by comparison with (2.17).

For a sample of observations on earnings, $y_1, y_2, ..., y_N$, maximum likelihood estimates are obtained by choosing the θ_is to maximise the log-likelihood function:

$$\ln L = \sum_{j=1}^{N} \sum_{i=1}^{M} \theta_i \psi_i(y_j) - N\eta \tag{2.25}$$

Standard iterative optimisation routines can be used to determine the θ_is which maximise (2.25). In maximising (2.25), the normalising constant η in (2.24) can be determined numerically using Gaussian quadrature routines.

Suppose instead that earnings data are available in the form of a frequency table. Lye and Martin (1994) introduced a least squares approach for estimating the parameters in (2.23) when the data are grouped, motivated by the least squares estimator of the Pareto distribution suggested by Johnson and Kotz (1970, p.235), and the work of Finch (1989). The distinguishing feature of the least squares estimator proposed by Lye and Martin (1994) is that it is based on frequencies. The error terms are therefore more likely to satisfy the assumption of independence than in the approach of Johnson and Kotz, which uses cumulative frequencies. Let the number of classes of equal width be K, with midpoints y_k, $k = 1, 2, ...K$, and define O_k as the observed frequency corresponding to the kth class interval. The expected frequency in the kth group, E_k, can be written as:

$$E_k = \phi^{-1} \exp \left[\sum_{i=1}^{M} \theta_i \psi_i(y_k) \right] \tag{2.26}$$

where ϕ is the normalising constant and values are assumed to be concentrated at the class midpoint.

The least squares estimator consists of defining a logarithmic linear regression relationship between the observed frequency O_k and the expected frequency E_k, as given by (2.23). Letting v_k be an error term, the regression equation is:

$$\ln(O_k) = \ln(E_k) + v_k \tag{2.27}$$

The ratio, O_k/E_k, is assumed to be lognormally distributed with unit mean, so that the estimates are constrained to be non-negative. Substituting (2.26) into (2.27) gives:

$$\ln(O_k) = -\ln \phi + \sum_{i=1}^{M} \theta_i \psi_i(y_k) + v_k \tag{2.28}$$

The parameters can be estimated using a standard ordinary least squares procedure. Since the sampling theory of the estimators has not been fully worked out, conventional formulae to compute standard errors and hence test statistics are not reported below. Instead, a goodness of fit measure based on a Chi-square statistic is calculated in order to compare models.

Denoting least squares estimators by ˆ, the estimate of the kth frequency is:

$$\hat{E}_k = \psi^{-1} \exp\left[\sum_{i=1}^{M} \hat{\theta}_i \psi_i(y_k)\right] \tag{2.29}$$

where:

$$\psi = \frac{1}{N} \sum_{k=1}^{K} \exp\left[\sum_{i=1}^{M} \hat{\theta}_i \psi_i(y_k)\right] \tag{2.30}$$

In the case where the required maximum likelihood software is not available, the least squares estimator discussed above can also be used by grouping the data into a frequency table. Since there is a loss of information from grouping, some loss of efficiency relative to the maximum likelihood estimator can be expected from using the least squares estimator. However, the Monte Carlo experiments performed in chapters 3 and 5 show that, provided the sample size is large, the least squares estimator performs very well.

2.2.2 The Conditional Distribution

In the estimation procedures discussed so far, individuals are assumed to be homogeneous. To allow for heterogeneity, assume that associated with the jth individual is a set of characteristics given by the vector z_j. In terms of the economic model presented above, this amounts to allowing the structural parameters to be functions of the variables in z. Examples of factors that would help determine the earnings of individuals are the level of education and experience.

Maximum likelihood parameter estimates of the heterogeneous earnings model when the data are ungrouped, are obtained by maximising the log likelihood function:

$$\ln L = \sum_{j=1}^{N} \sum_{i=1}^{M} \theta_i(z_j) \psi_i(y_j) - \sum_{j=1}^{N} \eta_j \tag{2.31}$$

where the normalising constant is:

$$\eta_j = \ln \int \exp \left[\sum_{i=1}^{M} \theta_i(z_j)\psi_i(y) \right] dy \qquad (2.32)$$

The main difference between (2.32) and (2.24) is that the former needs to be computed for each individual in the sample, at each iteration. Thus the algorithm is computationally burdensome, especially for large data sets. The adaptation of the least squares procedure to the conditional distribution context is discussed in chapter 5 and used in chapter 6.

2.3 Applications

This section applies the equilibrium earnings model to both unconditional and conditional distributions of earnings and the results are compared with existing distributional specifications. The data are from the 1987 March Supplement (Annual Demographic File) of the United States Current Population Survey (CPS) which contains information on individual earnings and labour market characteristics in 1986. The total sample is 4927 individuals. Since there is a bunching at the top end of the distribution caused by an upper limit on hours worked, 104 individuals are removed from the sample used in this chapter. The earnings variable is a measure of weekly earnings obtained as the product of hours per week and wages per hour.

2.3.1 Estimates of the Unconditional Earnings Distribution

Maximum likelihood estimates of the structural parameters for the unconditional generalised gamma earnings distribution in (2.13) are given in Table 2.1, along with estimates of the standard gamma distribution where $\theta_3 = \theta_4 = 0$. The asymptotic standard errors, shown in parentheses, are based on the inverse of the hessian matrix. In addition, the results are reported of the generalised lognormal distribution as given by (2.17) and the standard lognormal distribution where in (2.17) $\theta_1 = \theta_4 = 0$. Inspection of the standard errors in the case of the generalised gamma distribution shows that the estimates of θ_1 and θ_2 are statistically significantly different from zero, although those of θ_3 and θ_4 are not. More important, however, is an overall measure of the model's performance. This is achieved by the Akaike Information Criterion (AIC) reported in Table 2.1. This is computed as $AIC = -\ln L + K$, where K, the number of estimated parameters, acts as a penalty for the inclusion of additional parameters. The result in Table 2.1 shows that the generalized gamma distribution is preferred to the

Table 2.1: The unconditional earnings distribution: MLE

Parameter	Gen. Gamma	Gamma	Gen. Lognorm.	Lognorm.
θ_1	2.0298	2.4365	-0.1604	
	(0.1556)	(0.0698)	(0.1929)	
θ_2	-4.9953	-6.8078	-0.7728	-1.3882
	(1.0931)	(0.1471)	(1.3183)	(0.0361)
θ_3	-0.5772		1.1212	-3.2810
	(1.0888)		(3.2684)	(0.0586)
θ_4	-0.1813		-5.7674	
	(0.4079)		(3.3140)	
ln L	-39.9708	-50.3252	-39.1348	-318.9790
AIC	43.9709	52.3252	43.1348	320.9790

Table 2.2: The unconditional earnings distribution: least squares

Parameter	Gen. Gamma	Gamma	Gen. Lognorm.	Lognorm.
Constant	8.7108	5.1751	-9.4945	-3.2895
θ_1	5.6220	4.2579	-0.8121	
θ_2	-14.7650	-8.2186	-5.6642	-1.9856
θ_3	3.2468		-10.1307	-3.4505
θ_4	-0.3946		6.2802	
χ^2 stat.	45.0612	133.9298	46.4094	110.6682
d.o.f.	9	11	9	11

gamma distribution as the AIC is minimised for the generalised gamma distribution. Furthermore, inspection of the AIC for the generalised lognormal distribution shows that this distribution also performs very well; indeed, it is diffucult to distinguish between the two generalised forms.

The results using the least squares estimator are given in Table 2.2 for the same four types of distribution given in Table 2.1. The data are grouped into 21 cells with the first cell having a mid-point of \$100, the second a mid-point of \$200, and so on. The Chi-square statistics shown in Table 2.2 indicate that both of the generalised distributions perform significantly better than their standard forms even when the loss of degrees of freedom is taken into account. The superiority of the generalised distributions are further highlighted in Table 2.3, which gives the actual and expected frequencies from the alternative models.

Table 2.3: Comparison of actual and expected frequencies: least squares

Group midpoint	Actual	Gen. Gamma	Gamma	Gen. Lognorm.	Lognorm.
100	17	17	21	17	14
200	180	206	179	193	276
300	598	536	443	533	657
400	785	764	663	776	817
500	729	800	754	813	772
600	650	702	720	708	636
700	622	553	611	553	487
800	428	408	474	404	359
900	301	288	344	284	258
1000	162	198	237	194	183
1100	129	134	156	132	130
1200	77	90	99	89	91
1300	77	60	61	60	65
1400	44	41	37	40	46
1500	24	27	22	28	33

2.3.2 Estimates of the Conditional Earnings Distribution

In order to illustrate the way in which further heterogeneity may be introduced, suppose that the demand for labour is affected by the labour market experience, t, and the education of individuals, h, measured in terms of years of schooling. The constant term, δ, in the labour demand function can be specified as a function of these variables. This can be seen to imply that the terms θ_1 and θ_2 in the stationary distribution are functions of those variables. Suppose that, following the standard use of a quadratic experience profile, the relevant functions can be written as:

$$\theta_1 = \delta_0 + \delta_1 h_n + \delta_2 t_n + \delta_3 t_n^2$$

$$\theta_2 = \gamma_0 + \gamma_1 h_n + \gamma_2 t_n + \gamma_3 t_n^2$$

Maximum likelihood estimates of the conditional earnings model are given in Table 2.4. The asymptotic standard errors, shown in parentheses, are based on the inverse of the hessian matrix.

To highlight the properties of the estimated model, the effects of changes in education and experience on the earnings distribution are shown in Figure

Table 2.4: The conditional earnings distribution: MLE[(a)]

Variable	Parameter	Gen. Gamma
Constant	δ_0	1.6201
		(0.4669)
h	δ_1	-0.0019
		(0.0329)
t	δ_2	0.0846
		(0.0442)
t^2	δ_3	-0.0016
		(0.0006)
Constant	γ_0	-18.3953
		(1.4232)
h	γ_1	0.8770
		(0.0727)
t	γ_2	0.2878
		(0.0996)
t^2	γ_3	-0.0040
		(0.0015)
	θ_3	-3.0728
		(1.6999)
	θ_4	0.0570
		(0.4717)
	ln L = 775.5155	AIC = 795.5155

(a) Asymptotic standard errors in brackets

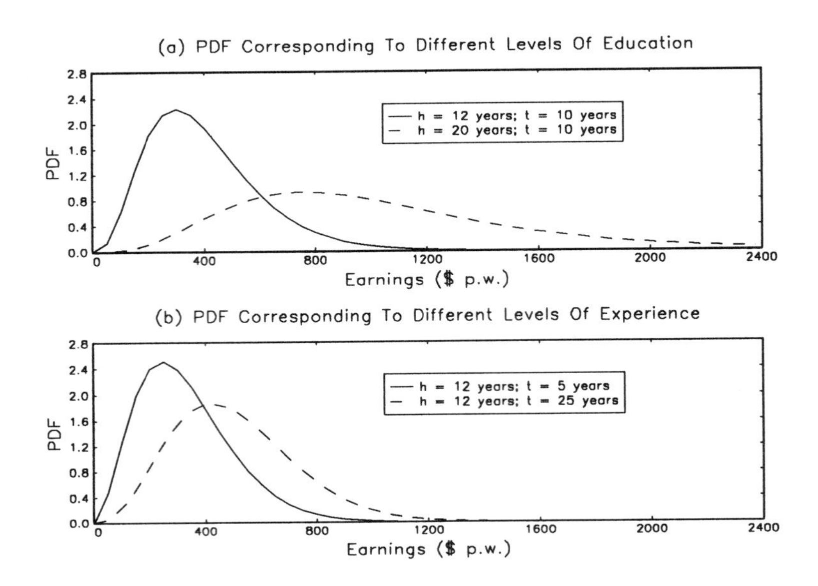

Figure 2.1: Earnings distributions: MLE

2.1. In both cases the predictions of the model agree with well-established results. Increases in both education and experience shift the earnings distribution to the right so that, on average, people with higher education and more experience have higher earnings levels. Furthermore, the dispersion increases with experience. In particular, an increase in education from 12 years to 20 years leads to an increase in the average earnings level from \$397 to \$1023, assuming that experience is 10 years. The results also show that for people with 12 years of education, a person with 25 years experience earns on average \$514, which is more than the amount of \$327 which would be earned on average by a person with only 5 years of experience. These figures compare with overall sample means of \$504 for weekly earnings, 13 years of education, and 19 years experience.

2.4 Conclusions

This chapter has explored the integration of a simple economic model of labour supply and demand with a stochastic process which causes shocks to the structural form of the model. For each individual, the number of hours of labour supplied was treated as the reciprocal supply associated with the demand for goods, or real income, while the demand for labour was a function of the real wage rate. Equilibrium earnings, the product of the wage rate and the number of hours worked, were derived as the root or roots of a cubic equation. An error correction process was then used to derive the stationary distribution associated with the model. This was found to be a generalised gamma distribution. An alternative specification of the structural form of the model produced a generalised lognormal distribution.

The generalised distributions have the advantage of nesting other special cases, such as the standard gamma or lognormal, which have been used in the statistical analysis of income distribution, depending on particular assumptions about the structural form. A special feature of these generalized distributions, in contrast with their standard forms, is that they can display multimodality corresponding to multiple equilibria in the structural form of the economic model. Furthermore, the parameters can be allowed to vary over time, so that the model has the potential to allow for macroeconomic changes to alter the form of the personal distribution of earnings. In some contexts the earnings distribution may be truncated and this can be handled using the methods of chapter 3.

It was possible to use estimation methods developed for the generalised exponential distribution, which encompasses generalisations of the gamma and lognormal distributions. These models were found to perform well using US data from the Current Population Survey in that they were able to capture the empirical characteristics of the earnings distribution better than the standard forms. The specification was extended to allow the structural parameters to depend on specified characteristics such as the experience and education of individuals. This provided a mechanism for understanding the effects of changes in both education and experience on the shape of the conditional earnings distribution.

Chapter 3

Truncated Distribution Families

Jenny N. Lye and Vance L. Martin

This chapter discusses the estimation of regression models when information on both the dependent and independent variables is sampled from a subset of a population. This is known as a truncated regression model and estimation procedures which do not take the sampling characteristics of the data into account give rise to biased estimates; see Greene (1990). This is typical of earnings functions which are estimated for families with incomes in a particular range; see Hausman and Wise (1977). Another example arises with endogenous stratification where the population is divided into a number of strata on the basis of income but one group is undersampled; see Maddala (1983, pp.171-172). This chapter concentrates on addressing problems that can arise in the area of income distribution; the approach has general applicability in areas where information is not complete. For example, insurance claims in excess of a particular value only may be reported; see, for example, Hogg and Klugman (1983) and chapter 7 below. Another example includes estimating exchange rate models when the exchange rate is constrained to operate within a region, that is, a target zone; see Peseran and Samiei (1992).

In estimating and testing truncated regression models, it is necessary to make assumptions about the underlying distribution. For many models used in applied work, the truncated normal distribution is commonly chosen. However, estimates based on this assumption may be poor when the true distribution contains non-normal characteristics such as skewness and or fat-tails. Some earlier work on generalising distributions in truncated regression models was carried out by Poirier (1978) who allowed for the possibility of skewness in the underlying variable before it was truncated by transforming it within the class of Box-Cox transformations, and by Amemiya and Boskin (1974) who used the lognormal distribution. An alternative approach which did not require knowledge of the true distribution was by Powell (1983) who proved the consistency and asymptotic normality of the least absolute deviations estimator.

This chapter uses the generalised exponential framework which is extended to distributions which are truncated from either above or below, or from both. The main result is that a number of commonly used truncated distributions can be viewed as subordinates of a generalisation of the beta distribution. These subordinates include, where there is double truncation, the truncated normal and exponential distributions, and the beta and power distributions. When there is single truncation, the subordinates include the gamma, lognormal and the Pareto distributions. Some generalisations of these existing distributions can be derived within the generalised beta class.

One advantage of this framework is that it extends the modelling framework of chapter 2 to the cases where variables are either truncated or censured. In doing so, a more formal theoretical mechanism for deriving alternative models of income distribution which have been used in empirical work is presented; see, for example, McDonald (1984). In particular, the chapter extends the work of Hausman and Wise (1977) by investigating the properties of a family of truncated distributions based on a generalisation of the beta distribution. Another advantage of this approach is that since it is possible to embed a range of competing distributions within the generalised beta distribution, classical hypothesis tests can be used to discriminate between them.

The generalised beta distribution is derived in Section 3.1 within a stochastic continuous time model, and its properties are given in Section 3.2, with attention paid to identifying the interrelationships that exist between its subordinates. Section 3.3 provides a maximum likelihood algorithm and a least squares estimator for estimating the parameters of the model; see chapter 5 for further discussion of the properties of these estimators. Estimation issues for regression models with errors distributed as generalised beta are discussed in Section 3.4. Lagrange multiplier tests of the distributional parameters are derived in Section 3.5 for truncated distributional models.

3.1 A Stochastic Model of Income

Consider the following Ito stochastic differential equation which represents the temporal movements in income; see also Creedy and Martin (1994b,c), and chapter 2 above:

$$dY_t = \mu(Y_t, X_t)dt + \sigma(Y_t, Z_t)dW_t \quad (3.1)$$

where Y_t is income, X_t represents a vector of explanatory variables which affect the mean of income, Z_t represents a vector of explanatory variables which affect the variance of income, and W_t is a Wiener process. The expressions $\mu(Y_t, X_t)$ and $\sigma^2(Y_t, Z_t)$ represent the mean and the variance of movements in income, dY, respectively. These are general expressions which are defined explicitly below.

The solution of the income distribution $f(y_t)$, is derived from the Kolmogorov forward equation:

$$\frac{\partial f}{\partial t} = -\frac{\partial(\mu f)}{\partial y} + \frac{\partial^2(\sigma^2 f)}{\partial y^2} \quad (3.2)$$

The stationary distribution, $f^*(y)$ is derived by setting $\partial f/\partial t = 0$ above. The solution is:

$$f^*(y) = exp\left[\int^y \left(\frac{2\mu(u) - d\sigma^2(u)/du}{\sigma^2(u)}\right)du - \eta\right] \quad (3.3)$$

where η is the normalizing constant such that $\int f^*(y)dy = 1$.

This equilibrium distribution is a general expression which depends upon the functions given for the mean μ and the variance σ^2. Suppose that the mean is a nonlinear function of income which can be approximated by the $(M-1)th$ order polynomial:

$$\mu(Y, X) = \sum_{i=0}^{M-1} \alpha_i(X)Y^i \quad (3.4)$$

where $\alpha_i(X)$ are general functions of the explanatory variables given in X. In choosing the specification for the variance, the truncation characteristics of the data need to be taken into account. In particular, if the data are truncated above and below at the points $Y^U$and Y^L respectively, then an appropriate specification for the variance is:

$$\sigma^2(Y) = \beta(Z)(Y - Y^L)(Y^U - Y) \quad (3.5)$$

where $\beta(Z) > 0$ is a general function of the explanatory variables in Z.

Given the expressions for the mean and variance in (3.4) and (3.5) respectively, the equilibrium income distribution can be derived explicitly. For example, letting $M = 4$, this distribution is:

$$f^*(y) = \exp\left[\theta_1 \ln(y - Y^L) + \theta_2 \ln(Y^U - y) + \theta_3 y + \theta_4 y^2 - \eta\right], \; Y^L < y < Y^U \quad (3.6)$$

This is a general expression which represents a generalised beta distribution. It also encompasses as special cases a number of other distributions which have been used to model the distribution of income. For example, letting $\theta_1 = \theta_2 = 0$ yields the truncated normal distribution which was used by Hausman and Wise (1977), whereas setting $\theta_1 = \theta_2 = \theta_4 = 0$ yields the truncated exponential distribution which has been used by Hogg and Klugman (1983). When there is single truncation from below such that $Y^L = 0$, the generalised beta distribution also contains as subordinates, both the Rayleigh ($\theta_2 = \theta_3 = 0$) and gamma ($\theta_2 = \theta_4 = 0$) distributions which have been used extensively in the income distribution literature; see, for example, McDonald (1984).

3.2 A Family of Truncated Distributions

3.2.1 Doubly Truncated Distributions

The model of income distribution discussed above resulted in a distribution referred to as a generalised beta distribution. The generalised beta distribution derived for the income distribution model was based on the assumption that income followed a stochastic differential equation. An alternative way to motivate the generalised beta distribution is to follow Cobb *et al.* (1983) and Lye and Martin (1993a) by generalising the Pearson differential equation. For example, let U be a random variable with support (A,B) which is governed by the differential equation:

$$\frac{df}{du} = \frac{-g(u)f(u)}{h(u)} \tag{3.7}$$

where $g(u) = \sum_{i=0}^{M-1} \alpha_i u^i$, and $h(u) = (u - A)(B - u)$. The solution of (3.7) yields the generalised beta distribution:

$$f(u) = exp\left[\theta_1 \ln(u - A) + \theta_2 \ln(B - u) + \sum_{i=3}^{M} \theta_i u^{i-2} - \eta\right], \quad A < u < B \tag{3.8}$$

where the normalising constant is given by:

$$\eta = \ln \int_A^B \exp\left[\theta_1 \ln(u - A) + \theta_2 \ln(B - u) + \sum_{i=3}^{M} \theta_i u^{i-2}\right] du \tag{3.9}$$

Some of the distributional properties of the following generalised beta distribution:

$$f(u) = \exp\left[\theta_1 \ln(1 + u) + \theta_2 \ln(1 - u) - 0.5u^2 - \eta\right] \tag{3.10}$$

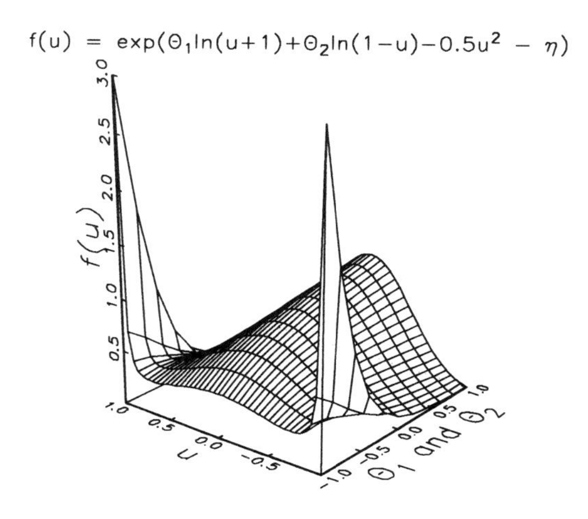

Figure 3.1: The generalised beta distribution

are highlighted by the surface plot in Figure 3.1 which displays a sequence of densities for alternative parameterisations. The truncation points in (3.8) are chosen as $A = -1$ and $B = 1$. For values of $\theta_1 = \theta_2 = 0$, the density is unimodal with the mode centred at zero. The density becomes more peaked as the values of θ_1 and θ_2 become positive. However, as θ_1 and θ_2 become negative, the distribution becomes trimodal with modes occurring at the end points as well as the interior mode at zero. As θ_1 and θ_2 become more negative, the interior mode reduces in height as more of the mass of the density function is located at the end points.

The generalised beta distribution constitutes an important family of doubly truncated distributions since it encompasses a number of truncated distributions, together with generalisations of these distributions, that are widely used in empirical work. Some examples are given below.

Example 1: The Truncated Normal Distribution
The truncated normal distribution represents one of the most widely used distributions in the area of qualitative response models. This distribution is derived from (3.8) by imposing the restrictions $\theta_1 = \theta_2 = \theta_3 = 0$, $\theta_4 = -1/2$, $\theta_i = 0$, $\forall i > 4$. The truncated normal distribution is:

$$f(u) = \exp\left[-u^2/2 - \eta\right], -A < u < B \tag{3.11}$$

where $\Phi(.)$ represents the cumulative normal distribution and the normalising constant is given by:

$$\begin{aligned}\eta &= \ln \int_A^B \exp\left[-u^2/2\right] du \\ &= \ln\left(\sqrt{2\pi}\right)\left[\Phi\left(B\right) - \Phi\left(A\right)\right]\end{aligned}$$

Example 2: The Generalised Truncated Normal Distribution

An appropriate generalisation of the truncated normal distribution is given by imposing the following restrictions on (3.8): $\theta_1 = \theta_2 = 0$, $\theta_i \neq 0$, $\forall i > 2$. The generalised normal truncated distribution is given by:

$$f(u) = \exp\left[\sum_{i=3}^{M} \theta_i u^{i-2} - \eta\right], \quad A < u < B \tag{3.12}$$

Example 3: The Beta Distribution

A special case of the generalised beta distribution is the beta distribution which is derived by imposing the restrictions $\theta_1, \theta_2 > -1$, $\theta_i = 0$, $\forall i > 2$. Letting A=0 and B=1, yields the standardized form of the beta distribution which is given by:

$$f(u) = exp\left[\theta_1 \ln(u) + \theta_2 \ln(1-u) - \eta\right], \quad 0 < u < 1 \tag{3.13}$$

where θ_1, $\theta_2 > -1$. Another special case of the generalized beta distribution is the generalized rho distribution when $A = -1$ and $B = 1$. This distribution was used by Lye and Martin (1993b) to approximate the distribution of the correlation coefficient in the spurious regression model.

Example 4: The Generalised Power Distribution

A power distribution is a special case of the beta distribution occurring when $\theta_2 = 0$ in (3.13). This suggests that a generalised power distribution can be defined by imposing this restriction in (3.8) together with $\theta_1 > -1$:

$$f(u) = exp\left[\theta_1 \ln(u-A) + \sum_{i=3}^{M} \theta_i u^{i-2} - \eta\right], \quad A < u < B \tag{3.14}$$

Example 5: The Truncated Exponential Distribution
A truncated exponential distribution can be defined by constraining all parameters in (3.8) to zero with the exception of θ_3, which is constrained to be negative. The distribution is:

$$f(u) = \exp\left[\theta_3 u - \eta\right], \quad A < u < B \tag{3.15}$$

where the normalising constant is given by:

$$\begin{aligned} \eta &= \ln \int_A^B \exp\left[\theta_3 u\right] du \\ &= \ln\left(\exp\left(\theta_3 B\right) - \exp\left(\theta_3 A\right)\right) - \ln \theta_3 \end{aligned}$$

The mean associated with (3.8) for alternative values of θ_3 are given in Figure 3.2 for the generalised beta ($\theta_1 = \theta_2 = 1$, $\theta_4 = -2$), normal ($\theta_1 = \theta_2 = 0$, $\theta_4 = -2$) and exponential ($\theta_1 = \theta_2 = \theta_4 = 0$) truncated distributions. The values of θ_3 range from -8 to 8, and the truncation points are $A = -1$ and $B = 1$. The results show that the relationship between the mean and θ_3 is sigmoidal and that for changes of θ_3 around the origin the mean is more responsive for the truncated exponential distribution and least for the generalised beta.

3.2.2 Singly Truncated Distributions

A set of distributions where truncation is one-sided, commonly known as singly truncated distributions, can be defined within the generalised framework discussed above. In those cases where truncation only occurs from below, the upper truncation point can be set to $B = \infty$. This suggests that $h(u)$ in (3.7) should be chosen as $h(u) = u - A$.

Example 6: The Generalised Half-Normal Distribution
A normal random variable U, which has support on $(0, \infty)$, has a half-normal distribution. This suggests that a generalised half-normal distribution can be defined by (3.12) with $A = 0$ and $B = \infty$.

Example 7: The Generalised Truncated Gamma Distribution
A special case of the power distribution is the gamma distribution which is derived by defining the support of the random variable to be (A, ∞) and

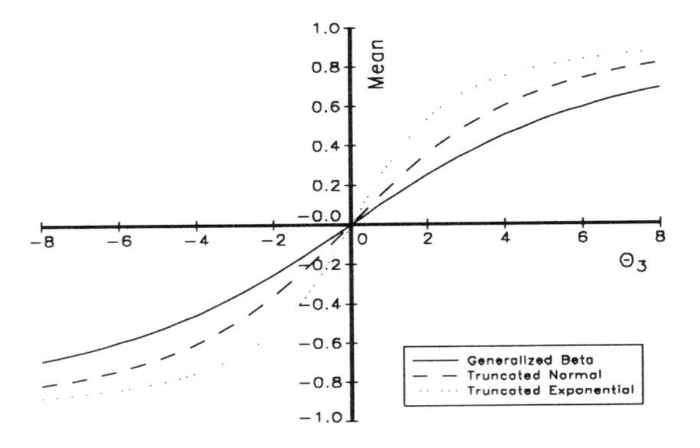

Figure 3.2: Mean of alternative generalised truncated distributions

setting $\theta_i = 0 \ \forall i > 3$ in (3.14). For many applications of the gamma distribution the lower truncation point is set to $A = 0$; see Johnson and Kotz (1970).

The generalised power distribution in (3.14) can also be interpreted as a generalised gamma distribution when the support of the random variable is chosen as (A, ∞). This distribution has been discussed by Cobb *et al.* (1983), Martin (1990) and Lye and Martin (1993a). It has been derived by Creedy and Martin (1994b) in an economic model of price distributions.

Example 8: The Generalised Pareto Distribution

The Pareto distribution was the first statistical model used to investigate the distribution of income; see Creedy (1985). This distribution can be derived from the power distribution with support $(0, B)$, by defining a new random variable as $V = U^{-1}$; see Johnson and Kotz (1970, p.247). The generalised Pareto distribution is derived from (3.14) by setting $A = 0$, and using the transformation of variable technique. This generalized Pareto distribution is given by:

$$f(v) = \exp\left[-(2+\theta_1)\ln(v) + \sum_{i=3}^{M} \theta_i v^{2-i} - \eta\right], \quad 1/B < v < \infty \qquad (3.16)$$

There exist other generalisations of the Pareto distribution which can be shown to be related to (3.16). For a discussion of these alternatives, see Johnson and Kotz (1970).

Finally, the standard Pareto distribution is given by (3.16) with $\theta_1 < -1$, $\theta_i = 0$, $\forall i \geq 3$, and $B > 0$.

Example 9: The Generalised Truncated Lognormal Distribution
A generalisation of the lognormal distribution was given by Lye and Martin (1993a). To extend this generalisation to the case where the distribution is truncated from below, for example, it is appropriate to redefine the expressions for $g(u)$ and $h(u)$ in (3.7), as:

$$\begin{aligned} g(u) &= \alpha_0 \ln(u-A) + \sum_{i=1}^{M-1} \alpha_1 u^{i-1} \\ h(u) &= u - A \end{aligned}$$

The solution of the differential equation is (see Lye and Martin, 1993a):

$$f(u) = \exp\left[\theta_1 (\ln(u-A))^2 + \theta_2 \ln(u-A) + \sum_{i=3}^{M} \theta_i u^{i-2} - \eta\right], \quad A < u < \infty \tag{3.17}$$

3.3 Estimation of Distributions

3.3.1 Maximum Likelihood

The parameters of the generalised beta distribution can be estimated by maximum likelihood procedures. For the sample, $u_1, u_2, ..., u_T$, of size T, the log of the likelihood for the generalised beta distribution in (3.8) is:

$$\begin{aligned} \ln L &= \sum_t \ln f(u_t) \\ &= \sum_t \left[\theta_1 \ln(u_t - A) + \theta_2 \ln(B - u_t) + \sum_{i=3}^{M} \theta_i u_t^{i-2} - \eta_t\right] \end{aligned} \tag{3.18}$$

where the normalising constant η_t is given by (3.9). For simplicity, the functions, $\theta_i, i = 1, 2, ..., M$, are treated as constant and not functions of a set of explanatory variables. This assumption means that the normalising constant is invariant with respect to t.

When the truncation points, A and B, are known, (3.18) can be maximised by using standard iterative procedures such as Berndt *et al.* (1974) or Newton-Raphson. The normalising constant in (3.9) can be computed by using numerical integration techniques. Standard numerical software libraries, such as GAUSS, can be used to solve these problems.

When the truncation points are unknown, (3.18) needs to be maximised with respect to both the distributional parameters θ_1, $i = 1, 2, ..., M$, and the truncation parameters A and B. This problem is akin to deriving consistent maximum likelihood estimators of the parameters in the Pareto distribution (see Quandt, 1966) as the likelihood function is unbounded with respect to the support parameter(s). The solution is obtained by maximising the likelihood function with respect to the support parameters subject to the constraint that the data are bounded by these parameters. This leads to the support parameter estimators being a function of order statistics. In particular, consistent estimators can be obtained by setting the lower truncation point equal to the first order statistic and the upper truncation point equal to the largest order statistic. Thus, for the generalised beta distribution when the truncation points are unknown, the approach adopted is to choose $\hat{A} = \min\{u_1, u_2, ..., u_T\}$, $\hat{B} = \max\{u_1, u_2, ..., u_T\}$, and maximise (3.18) with respect to the parameters, $\theta_i, i = 1, 2, ..., M$, by using standard optimising procedures.

3.3.2 Least Squares

For problems where the data are grouped into T frequencies, $t = 1, 2, ..., T$, the parameters can be estimated by least squares. Maximum likelihood and minimum chi-square estimators can also be derived for this class of problems, which are discussed in Lye and Martin (1993a), and chapters 2, 5 and 6. One advantage of the least squares approach over the maximum likelihood approach is that it is computationally more simple as it avoids the need to use numerical integration routines to compute the normalising constant. The approach adopted here is partly motivated by the least squares estimators of the Pareto distribution suggested by Johnson and Kotz (1970, p.235), and partly by the work of Finch (1989) and Martin (1990).

Assuming that the truncation parameters are known, the least squares estimators of θ_i, $i = 1, 2, ..., M$, can be derived by defining the relationship between E_t, the expected relative frequency of the tth group and O_t, the corresponding observed relative frequency, as:

$$O_t = E_t \ \exp(v_t) \tag{3.19}$$

where v_t is an error term. If E_t is given by the generalised beta distribution in (3.8), (3.19) can be written as:

$$\ln(O_t) = \theta_0 + \theta_1 \ln(u_t - A) + \theta_2 \ln(B - u_t) + \sum_{i=3}^{M} \theta_1 u_t^{i-2} + v_t \tag{3.20}$$

where $\theta_0 = \eta$, and is effectively a representative value such as the midpoint from the tth class interval. This represents a least squares regression equation where the parameters can be chosen to minimise $\sum v_t^2$.

One advantage of this approach is that the estimates of the frequencies are constrained to be non-negative. For example, letting $\widetilde{\theta}$ denote the least squares estimator of θ, the relative frequencies can be estimated as:

$$\widetilde{E}_t = \phi^{-1} \exp\left[\widetilde{\theta}_0 + \widetilde{\theta}_1 \ln(u_t - A) + \widetilde{\theta}_2 \ln(B - u_t) + \sum_{i=3}^{M} \widetilde{\theta}_i u_t^{i-2}\right] \tag{3.21}$$

where:

$$\phi = \sum_{t=1}^{T} \exp\left[\widetilde{\theta}_0 + \widetilde{\theta}_1 \ln(u_t - A) + \widetilde{\theta}_2 \ln(B - u_t) + \sum_{i=3}^{M} \widetilde{\theta}_i u_t^{i-2}\right]$$

In the case where the truncation points are unknown, the parameters $\{A, B, \theta_1, \theta_2, ..., \theta_M\}$ can be estimated by standard non-linear maximum likelihood techniques by assuming that v_t is normally distributed. This highlights a further advantage from using frequency functions; namely, the error terms v_t in (3.20), are more likely to be independent when based on frequencies than cumulative frequencies. For example, in estimating the parameters of the Pareto distribution by maximum likelihood with grouped data, the dependent variable is often expressed as a function of cumulative frequencies (see Johnson and Kotz, 1970, p.235; and Creedy, 1985, pp.27-28).

3.3.3 Monte Carlo Simulations

This section examines the small sample properties of both the maximum likelihood and least squares estimators using Monte Carlo experimentation. In each experiment the results reported are based on 1000 replications. Three population distributions are considered. The first is the following beta distribution:

$$f(u) = \exp[2.0 \ln(u) + 2.0 \ln(1.0 - u) - \eta] \tag{3.22}$$

where $0 \leq u \leq 1$ and:

$$\eta = \ln \int_0^1 \exp\left[2.0\ln(u) + 2.0\ln(1.0-u)\right] du$$

The second is the lognormal distribution:

$$f(y) = \exp\left[-0.78\ln(u)^2 + 2.125\ln(u) - \eta\right] \tag{3.23}$$

where $0 \leq u \leq \infty$ and:

$$\eta = \ln \int_0^\infty \exp\left[-0.78\ln(u)^2 + 2.125\ln(u)\right] du$$

The third distribution is a bimodal generalised beta distribution of the form:

$$f(u) = \exp\left[0.7\ln(u+1.5) + 0.5\ln(2.0-u) - 0.5u + 1.5u^2 - 0.15u^4 - \eta\right] \tag{3.24}$$

where $-1.5 \leq u \leq 2.0$ and:

$$\eta = \ln \int_{1.5}^{2} \exp\left[0.7\ln(u+1.5) + 0.5\ln(2.0-u) - 0.5u + 1.5u^2 - 0.15u^4\right] du$$

Beta random variables are generated using the relationship:

$$R = \frac{X_1}{X_1 + X_2} \tag{3.25}$$

where X_1 and X_2 are independent gamma variables which, in turn, are generated as $-[\ln(U_1) + \ln(U_2)]$, where U_1 and U_2 are independent uniform random variables over the range $(0, 1)$. Lognormal random variates, L, are generated using $L = \exp(N)$, where N is normally distributed with mean 2 and standard deviation 0.8. Generalised beta variates are generated using the inverse cumulative density function technique. Random uniform and normal variates are generated using the GAUSS subroutine RNDUS and RNDNS respectively, with the seed being set equal to 1061308800.

Case 1: Least Squares Estimator

Samples of size 500 and 1000 are generated using the methods defined above and the random numbers are then grouped into 20 frequency classes using

Table 3.1: Simulation results: least squares estimation

Distribution	True	T=500		T=1000	
		Mean	RMSE	Mean	RMSE
Beta	2.0	1.92	0.13	1.93	0.03
	2.0	1.91	0.14	1.92	0.04
Lognormal	-0.78	-0.80	0.10	-0.79	0.01
	2.13	2.29	1.02	2.13	0.09
Gen. Beta	0.70	1.08	0.23	1.56	0.03
	0.50	2.05	15.98	0.32	0.22
	-0.50	-0.19	1.32	-1.45	0.06
	1.50	1.59	0.03	1.70	0.04
	-0.15	-0.12	0.01	-0.16	0.01

the HIST command in GAUSS. In the case of the lognormal distribution, the particular distribution chosen has a very long tail so that generating extreme observations is possible. In such cases the extreme observations tended to influence the groupings of the frequency classes and hence the sample estimates. It was found that the influence of these outliers could be removed by deleting them from the sample. The outliers constituted at most 5 per cent of the sample.

The mean and root mean square error (RMSE) from the simulation experiments are reported in Table 3.1. The associated estimated distributions based on the mean values are illustrated and compared with the population distribution in Figure 3.3. Overall these results indicate that the least squares estimator performs well.

Case 2: Maximum Likelihood Estimator

Samples of size 50 and 1000 are generated using the methods defined above. The Newton-Raphson algorithm in the MAXLIK GAUSS library is used to estimate the parameters of each of the population densities. The mean and root mean square error (RMSE) from the simulation experiments are reported in Table 3.2. The associated estimated distributions based on the mean values are illustrated and compared with the population distribution in Figure 3.3. The results reported indicate that the maximum likelihood estimator performs well even for samples as small as 50.

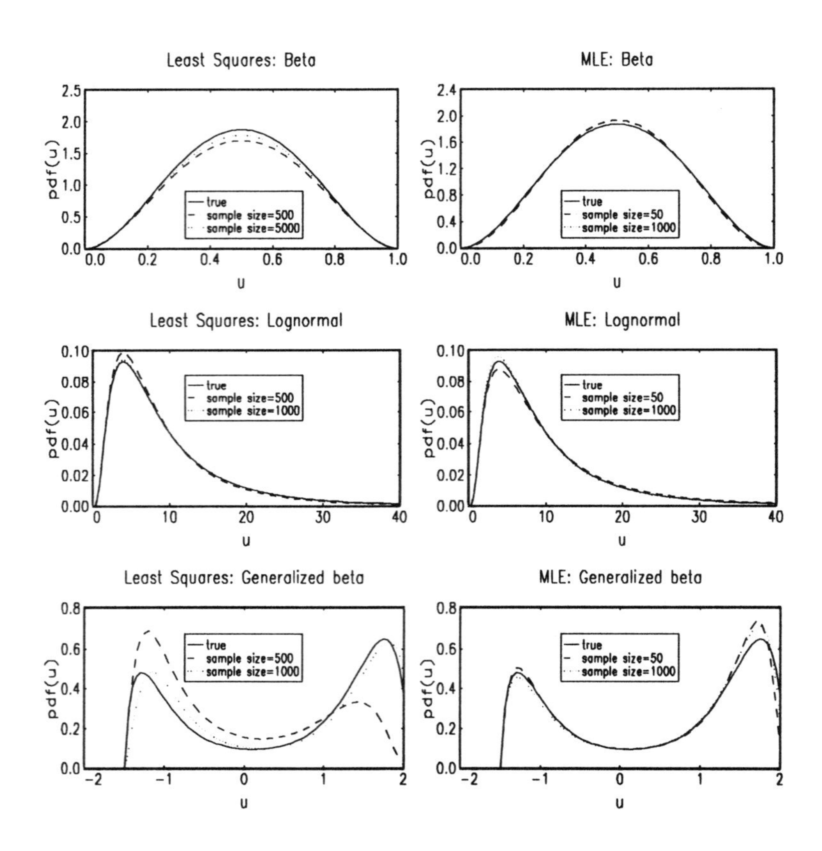

Figure 3.3: Monte Carlo properties of least squares and MLE estimators

Table 3.2: Simulation results: MLE

Distribution	True	T=50		T=1000	
		Mean	RMSE	Mean	RMSE
Beta	2.0	2.18	0.44	2.01	0.02
	2.0	2.18	0.40	2.01	0.02
Lognormal	-0.78	-0.72	0.03	-0.79	0.01
	2.13	1.96	0.49	2.10	0.17
Gen. Beta	0.70	1.02	0.47	0.79	0.01
	0.50	1.86	3.15	1.16	0.06
	-0.50	-0.03	1.01	-0.20	0.03
	1.50	1.56	0.39	1.50	0.02
	-0.15	-0.001	0.08	-0.08	0.01

3.4 Estimation of Regression Models

The discussion so far has concentrated on estimating distributional parameters when there are no explanatory variables. This situation is appropriate, for example, when data are only available on income and the aim is to model the empirical distribution. For other applications, such as models of earnings, there exists a set of characteristic variables such as experience and education, which affect the conditional mean of earnings. To allow for a set of explanatory variables, $\{X_{1t}, X_{2t}, ..., X_{Kt}\}$, in the estimation procedures discussed above, the approach adopted is to define the regression equation for the tth observation as:

$$Y_t = \lambda_t + \nu u_t \tag{3.26}$$

where:

$$\lambda_t = \beta_0 + \sum_{j=1}^{K} \beta_j X_{jt} \tag{3.27}$$

u_t is distributed according to the generalised beta distribution in (3.8), and Y_t is the dependent variable which has mean and standard deviation λ_t and ν respectively, and lower and upper truncation points, Y^L and Y^U, respectively.

Maximum likelihood estimation can proceed as above with the log likelihood given by:

$$\ln L = \sum_t \ln f(Y_t)$$

$$= \sum_t [\theta_1 \ln\left(Y_t - \lambda_t\right)/\nu - A_t) + \theta_2 \ln(B_t - (Y_t - \lambda_t)/\nu)$$
$$+ \sum_{i=3}^{M} \theta_i \left((Y_t - \lambda_t)/\nu\right)^{i-2} - \ln(\nu) - \eta_t] \qquad (3.28)$$

where the normalising constant is:

$$\eta_t = \ln \int_{A_t}^{B_t} exp\left[\theta_1 \ln(u - A_t) + \theta_2 \ln(B_t - u) + \sum_{i=3}^{M} \theta_i u^{i-2}\right] du \qquad (3.29)$$

and:

$$B_t = \frac{Y^U - \lambda_t}{\nu} \qquad A_t = \frac{Y^L - \lambda_t}{\nu} \qquad (3.30)$$

It is not possible to estimate all of the parameters in (3.28) since not all parameters are identified. One approach is to constrain θ_M to be equal to some constant. Note that to estimate the parameters of this model by maximum likelihood is computationally more burdensome than it is to estimate the model in subsection 3.3.1, since the numerical integrations in (3.29) have to be computed at each observation as the range of integration is a function of the data through the term λ_t.

Estimation of the model when the data are grouped into relative frequencies can proceed as in Section 4.2 by using (3.26) to substitute out u_t in (3.20). The estimating equation becomes:

$$\begin{aligned} \ln(O_t) &= \theta_0 + \theta_t \ln\left(\left(Y_t - \lambda_t\right)/\nu - A_t\right) + \theta_2 \ln\left(B_t - \left(Y_t - \lambda_t\right)/\nu\right) \\ &+ \sum_{i=3}^{m} \theta_i \left(\left(Y_t - \lambda_t\right)/\nu\right)^{i-2} + v_t \end{aligned} \qquad (3.31)$$

where λ_t is given by (3.27). This equation can be estimated by using standard iterative maximum likelihood procedures by assuming that v_t is normally distributed, and that θ_M equals a constant for identifiability.

3.5 Testing

A common assumption used in modelling and testing truncated samples is that the underlying distribution is truncated normal. However, the consequences of violating this assumption can be quite severe since in this case the maximum likelihood estimators can be inconsistent; see, for example, Bera *et al.* (1984). This result gives sufficient reason for testing the truncated normal distribution.

This section develops two Lagrange multiplier tests which test a sequence of hypotheses within the generalised beta distribution:

$$f = \exp\left[\theta_1 \ln(u_t - A_t) + \theta_2 \ln(B_t - u_t) + \theta_3 u_t + \theta_4 u_t^2 + \theta_5 u_t^3 + \theta_6 u_t^4 - \eta_t\right] \quad (3.32)$$

where $A_t < u_t < B_t$ and the normalising constant is:

$$\eta_t = \ln \int_{A_t}^{B_t} exp\left[\theta_1 \ln(u - A_t) + \theta_2 \ln(B_t - u) + \theta_3 u + \theta_4 u^2 + \theta_5 u^3 + \theta_6 u^4\right] du$$

where $u_t = (Y_t - \lambda_t)/\nu$ is a regression disturbance defined with zero mean and unit variance, A_t and B_t are defined as in (3.30), and η_t is the normalising constant.

In the first Lagrange multiplier test the joint hypothesis $\theta_1 = \theta_2 = 0$ is tested. If this hypothesis is rejected so that $\theta_1, \theta_2 \neq 0$, then the underlying distribution cannot be modelled by a truncated normal distribution. If, on the other hand, the null hypothesis is not rejected, that is, $\theta_1 = \theta_2 = 0$, then the underlying distribution is still being represented by a nonnormal truncated distribution in particular the generalised truncated normal distribution given by:

$$f = \exp\left[\theta_3 u_t + \theta_4 u_t^2 + \theta_5 u_t^3 + \theta_6 u_t^4 - \widetilde{\eta}_t\right]$$

where the normalising constant is now:

$$\widetilde{\eta}_t = \ln \int_{A_t}^{B_t} \exp\left[\theta_3 u + \theta_4 u^2 + \theta_5 u^3 + \theta_6 u^4\right] du$$

To test the suitability of the truncated normal distribution as the underlying distribution, a second Lagrange multiplier test can be performed which has as its null hypothesis, $\theta_5 = \theta_6 = 0$.

The Lagrange multiplier test of $\theta_1 = \theta_2 = 0$, is derived from the log likelihood function as follows. For a sample of size $t = 1, 2, ..., T$, the negative of the log likelihood is:

$$\begin{aligned} -\ln L &= \sum_t \eta_t - \theta_1 \sum_t \ln(u_t - A_t) - \theta_2 \sum_t \ln(B_t - u_t) \\ &\quad -\theta_3 \sum_t u_t - \theta_4 \sum_t u_t^2 - \theta_5 \sum_t u_t^3 - \theta_6 \sum_t u_t^4 \end{aligned}$$

The first order derivatives are:

$$\frac{-\partial \ln L}{\partial \theta_1} = \sum_t \frac{\partial \eta_t}{\partial \theta_1} - \sum_t \ln(u_t - A_t)$$
$$\frac{-\partial \ln L}{\partial \theta_2} = \sum_t \frac{\partial \eta_t}{\partial \theta_2} - \sum_t \ln(B_t - u_t)$$
$$\frac{-\partial \ln L}{\partial \theta_i} = \sum_t \frac{\partial \eta_t}{\partial \theta_i} - \sum_t u_t^{i-2}, \; i = 3, 4, 5, 6$$

which from the definition of η_t can be written as

$$\begin{aligned}\frac{-\partial \ln L}{\partial \theta_1} &= \sum_t \int_{A_t}^{B_t} \ln(u - A_t) \exp[\theta_1 \ln(u - A_t) + \theta_2 \ln(B_t - u) \\ &\quad + \theta_3 u + \theta_4 u^2 + \theta_5 u^3 + \theta_6 u^4 - \eta_t] du - \sum_t \ln(u_t - A_t)\end{aligned}$$
$$\begin{aligned}\frac{-\partial \ln L}{\partial \theta_2} &= \sum_t \int_{A_t}^{B_t} 1n(B_t - u) \exp[\theta_1 1n(u - A_t) + \theta_2 1n(B_t - u) \\ &\quad + \theta_3 u + \theta_4 u^2 + \theta_5 u^3 + \theta_6 u^4 - \eta_t] du - \sum_t \ln(B_t - u_t)\end{aligned}$$
$$\begin{aligned}\frac{-\partial \ln L}{\partial \theta_i} &= \sum_t \int_{A_t}^{B_t} u^{i-2} \exp[\theta_1 \ln(u - A_t) + \theta_2 \ln(B_t - u) \\ &\quad + \theta_3 u + \theta_4 u^2 + \theta_5 u^3 + \theta_6 u^4 - \eta_t] du - \sum_t u_t^{i-2}\end{aligned}$$

for $i = 3, 4, 5, 6$. Evaluating the derivatives under the null ($\theta_1 = \theta_2 = 0$) gives

$$\begin{aligned}\left.\frac{-\partial \ln L}{\partial \theta_1}\right|_{\theta_1=\theta_2=0} &= \sum_t \int_{A_t}^{B_t} \ln(u - A_t) \\ &\quad \exp[\theta_3 u + \theta_4 u^2 + \theta_5 u^3 + \theta_6 u^4 - \widetilde{\eta}_t] du - \sum_t \ln(u_t - A_t)\end{aligned}$$
$$\begin{aligned}\left.\frac{-\partial \ln L}{\partial \theta_2}\right|_{\theta_1=\theta_2=0} &= \sum_t \int_{A_t}^{B_t} \ln(B_t - u) \\ &\quad \exp[\theta_3 u + \theta_4 u^2 + \theta_5 u^3 + \theta_6 u^4 - \widetilde{\eta}_t] du - \sum_t \ln(B_t - u)\end{aligned}$$
$$\left.\frac{-\partial \ln L}{\partial \theta_i}\right|_{\theta_1=\theta_2=0} = \sum_t E[u_t^{i-2}] - \sum_t u_t^{i-2}, \; i = 3, 4, 5, 6$$

where:

$$\widetilde{\eta}_t = \ln \int \exp\left[\theta_3 u + \theta_4 u^2 + \theta_5 u^3 + \theta_6 u^4\right] du$$

and:

$$E[u_t^{i-2}] = \int_{A_t}^{B_t} u^{i-2} \exp[\theta_3 u + \theta_4 u^2 + \theta_5 u^3 + \theta_6 u^4 - \widetilde{\eta}_t] du$$

The vector of first order derivatives when evaluated under the null is denoted by D and is equal to:

$$D = \left[\left.\frac{-\partial \ln L}{\partial \theta_1}\right|_{\theta_1=\theta_2=0}, \ldots, \left.\frac{-\partial \ln L}{\partial \theta_6}\right|_{\theta_1=\theta_2=0} \right]$$

These first order derivatives can be evaluated numerically. The second order derivatives which are also evaluated numerically are:

$$\frac{-\partial^2 \ln L}{\partial \theta_i \partial \theta_j} = \sum_t \frac{\partial^2 \eta}{\partial \theta_i \partial \theta_j}, \; i,j = 1, \ldots, 6$$

which when evaluated under the null become:

$$\left.\frac{-\partial^2 \ln L}{\partial \theta_1 \partial \theta_1}\right|_{\theta_1=\theta_2=0} = \sum_t \int_{A_t}^{B_t} \ln(u - A_t)^2 \exp[\theta_3 u + \theta_4 u^2 + \theta_5 u^3 + \theta_6 u^4 - \widetilde{\eta}_t] du - \sum_t \frac{\partial \eta_t}{\partial \theta_1} \frac{\partial \eta_t}{\partial \theta_1}$$

$$\left.\frac{-\partial^2 \ln L}{\partial \theta_1 \partial \theta_2}\right|_{\theta_1=\theta_2=0} = \sum_t \int_{A_t}^{B_t} \ln(u - A_t) \ln(B_t - u) \exp[\theta_3 u + \theta_4 u^2 + \theta_5 u^3 + \theta_6 u^4 - \widetilde{\eta}_t] du - \sum_t \frac{\partial \eta_t}{\partial \theta_1} \frac{\partial \eta_t}{\partial \theta_2}$$

$$\left.\frac{-\partial^2 \ln L}{\partial \theta_1 \partial \theta_j}\right|_{\theta_1=\theta_2=0} = \sum_t \int_{A_t}^{B_t} \ln(u - A_t) u^{j-2} \exp[\theta_3 u + \theta_4 u^2 + \theta_5 u^3 + \theta_6 u^4 - \widetilde{\eta}_t] du - \sum_t \frac{\partial \eta_t}{\partial \theta_1} \frac{\partial \eta_t}{\partial \theta_j}$$

$$\left.\frac{-\partial^2 \ln L}{\partial \theta_2 \partial \theta_2}\right|_{\theta_1=\theta_2=0} = \sum_t \int_{A_t}^{B_t} \ln(u - B_t)^2 \exp[\theta_3 u + \theta_4 u^2 + \theta_5 u^3 + \theta_6 u^4 - \widetilde{\eta}_t] du$$

$$- \sum_t \frac{\partial \eta_t}{\partial \theta_2} \frac{\partial \eta_t}{\partial \theta_2}$$

$$\left. \frac{-\partial^2 \ln L}{\partial \theta_2 \partial \theta_j} \right|_{H_0} = \sum_t \int_{A_t}^{B_t} \ln(u - B_t) u^{j-2}$$

$$\exp[\theta_3 u + \theta_4 u^2 + \theta_5 u^3 + \theta_6 u^4 - \widetilde{\eta}_t] du - \sum_t \frac{\partial \eta_t}{\partial \theta_2} \frac{\partial \eta_t}{\partial \theta_j}$$

$$\left. \frac{-\partial^2 \ln L}{\partial \theta_i \partial \theta_j} \right|_{H_0} = \sum_t \int_{A_t}^{B_t} u^{i+j-4} \exp[\theta_3 u + \theta_4 u^2 + \theta_5 u^3 + \theta_6 u^4 - \widetilde{\eta}_t] du$$

$$- \sum_t \frac{\partial \eta_t}{\partial \theta_i} \frac{\partial \eta_t}{\partial \theta_j}$$

for $j = 3, 4, 5, 6$.

The Lagrange multiplier test is given by Harvey (1990):

$$LM = D' H^{-1} D \tag{3.33}$$

where H is the matrix of second order derivatives evaluated under the null. In the case where the truncation points are known, Bera *et al.* (1984) show that the Lagrange multiplier statistic is distributed as χ_2^2 under H_0. In the case where the truncation points are unknown, provided that they can be consistently estimated, for example, by using the methods suggested above, the conditions for LM in (3.33) to be distributed as χ_2^2 given in Bera *et al.* still hold.

If the hypothesis $\theta_1 = \theta_2 = 0$ is not rejected, the next stage is to test the appropriateness of the truncated normal distribution. To do this a Lagrange multiplier test of $\theta_5 = \theta_6 = 0$ is derived. The test is of the same form as (3.33). To derive the test statistic, the negative of the log likelihood is now for a sample of size $t = 1, 2, ..., T$, defined as:

$$\ln L = \sum_t \widetilde{\eta}_t - \theta_3 \sum_t u_t - \theta_4 \sum_t u_t^2 - \theta_5 \sum_t u_t^3 - \theta_6 \sum_t u_t^4$$

The first order derivatives are:

$$\frac{-\partial \ln L}{\partial \theta_i} = \sum_t \frac{\partial \eta_t}{\partial \theta_i} - \sum_t u_t^{i-2}, \; i = 3, 4, 5, 6$$

which from the definition of η_t, can be written as:

$$\frac{-\partial \ln L}{\partial \theta_i} = \sum_t \int_{A_t}^{B_t} u^{i-2} \exp[\theta_3 u + \theta_4 u^2 + \theta_5 u^3 + \theta_6 u^4 - \widetilde{\eta}_t] du - \sum_t u_t^{i-2}$$

for $j = 3, 4, 5, 6$.

Evaluating the derivatives under the null $\theta_5 = \theta_6 = 0$ gives:

$$\left.\frac{-\partial \ln L}{\partial \theta_i}\right|_{\theta_5=\theta_6=0} = \sum_t E[u_t^{i-2}] - \sum_t u_t^{i-2}, \ i = 3, 4, 5, 6 \tag{3.34}$$

where:

$$E[u_t^{i-2}] = \int_{A_t}^{B_t} u^{i-2} \exp\left[\theta_3 u^3 + \theta_4 u^4 - \widehat{\eta}_t\right] du$$

and:

$$\widehat{\eta}_t = \ln \int \exp\left[\theta_3 u + \theta_4 u^2\right] du$$

The vector of first order derivatives when evaluated under the null is:

$$D = \left[\left.\frac{-\partial \ln L}{\partial \theta_3}\right|_{\theta_5=\theta_6=0}, \ldots, \left.\frac{-\partial \ln L}{\partial \theta_6}\right|_{\theta_5=\theta_6=0} \right]$$

These first order derivatives correspond to the first four moments of the truncated normal distribution; see Johnson and Kotz (1970).

The elements of H, the matrix of second order derivatives, are:

$$\frac{-\partial^2 \ln L}{\partial \theta_i \partial \theta_j} = \sum_t \frac{\partial^2 \eta}{\partial \theta_i \partial \theta_j}, \ i, j = 3, \ldots, 6$$

which when evaluated under the null become:

$$\left.\frac{-\partial^2 \ln L}{\partial \theta_i \partial \theta_j}\right|_{\theta_5=\theta_6=0} = \sum_t \int_{A_t}^{B_t} u^{(i+j-4)} \exp\left[\theta_3 u + \theta_4 u^2 + \theta_5 u^3 + \theta_6 u^4 - \overline{\eta}_t\right] du - \sum_t \frac{\partial \eta_t}{\partial \theta_i} \frac{\partial \eta_t}{\partial \theta_j}$$

for $i, j = 3, 4, 5, 6$, which is equal to:

$$\left.\frac{-\partial^2 \ln L}{\partial \theta_i \partial \theta_j}\right|_{\theta_5=\theta_6=0} = \sum_t \left[E[u_t^{i+j-4}] - E[u_t^{i-2}]E[u_t^{j-2}]\right], \ i, j = 3, 4, 5, 6 \tag{3.35}$$

where the expectations are with respect to the truncated normal distribution.

3.6 Conclusions

This chapter introduced a flexible class of distributions which are useful for modelling non-normal properties of the data in both doubly and singly truncated samples. Particular attention was given to investigating the properties of a generalisation of the beta distribution. Subordinates of this family for doubly truncated samples include the truncated normal, truncated exponential, power and beta distributions. Examples where only one side of the distribution is truncated are the gamma, Pareto and lognormal distributions and their respective generalisations. Two estimation methods were presented, including maximum likelihood and a method which only requires least squares techniques, and their small and large sample properties were investigated with the aid of Monte Carlo experimentation.

Chapter 4

Betit: A Flexible Binary Choice Model

Alex Bakker and Vance L. Martin

Binary models arise when the dependent variable takes on the discrete values, 0 and 1; see Domenich and McFadden (1975) and Amemiya (1981). The two most widely used specifications are the probit and logit models. In the probit model the underlying distribution is assumed to be normal, whereas for the logit model it is logistic; see Amemiya (1981) for a survey of the qualitative response literature and Maddala (1983) for fuller treatments of binary models. If these assumptions are violated it is well known that the parameter estimates are biased and inconsistent; see, for example Gabler *et al.* (1993), who present Monte Carlo evidence of the size of the bias for certain sample designs.[1]

This chapter introduces a general class of models which is flexible enough to allow for departures from the assumptions underlying the probit and logit models. The approach assumes that the probability of choosing an action is the mode of a generalized beta distribution; see Cobb *et al.* (1983) and Martin (1990) for a discussion of the generalised beta distribution. In keeping with existing terminology, this class of models is referred to as 'betit'.

An important feature of the betit model is that it can generate richer response surfaces than either the probit or the logit model. For the probit and logit models, the response surfaces are monotonic, continuous and symmetric about an index value of zero. These properties arise from using a cumulative density function which maps the index on to the unit interval thereby yielding probabilities between zero and one. This contrasts with the

[1]Ruud (1983, 1986) provides some counter examples where the probit model can yield consistent estimates when the underlying distribution is non-normal.

betit model where probability estimates are obtained by choosing the mode from a generalised beta distribution, which by construction lies in the unit interval.

The betit framework has several advantages. It can be parameterized to encompass the probit and logit models as well as be shown to be related to the class of heterogeneity models formulated by Heckman and Willis (1977). It is also related to the earlier generalisations suggested by Prentice (1976) and Copenhaver and Mielke (1977) as well as the semi-nonparametric class of models introduced by Gabler *et al.* (1993), which, in turn, is related to the semi-parametric approaches of Cosslett (1983) and Klein and Spady (1987). A pseudo maximum likelihood estimator is proposed which is asymptotically normally distributed under standard regularity conditions. Tests of normality are derived which are related to the Lagrange multiplier testing framework of Bera *et al.* (1984) as both are based on the Pearson system of distributions; see also Pagan and Vella (1989).

The traditional formulation of the binary choice problem is presented in Section 4.1. Section 4.2 introduces the betit model, its parameterization and relationship to logit and probit models. A pseudo maximum likelihood estimator of the betit model is outlined in Section 4.3 while alternative tests of normality are derived in Section 4.4. Section 4.5 examines the relative performance of logit, probit and betit models using married female labour force participation data. The betit model is found to be superior for a range of goodness-of-fit criteria and model specification tests to the logit and probit specifications.

4.1 The Binary Choice Problem

Let X_i represent a set of K attributes of the ith individual. The decision process is modelled as the relationship between an unobserved threshold level I_i^*, and an index of the attributes $I_i = X_i\beta$, where β is a set of K parameters. Formally the model is:

$$I_i^* = I_i + \varepsilon_i, \qquad i = 1, 2, \ldots, N \tag{4.1}$$

where ε_i represents an error term. Which alternative is chosen depends upon the relationship between the index and threshold level such that:

$$\begin{aligned} I_i^* &> 0 \Rightarrow Y_i = 1 \\ I_i^* &\leq 0 \Rightarrow Y_i = 0 \end{aligned} \tag{4.2}$$

where Y_i is a binary variable signifying whether the threshold is satisfied.

Defining π_i as the probability that an individual chooses $Y_i = 1$, then:

$$\begin{aligned} \pi_i &= \Pr(Y_i = 1) \\ 1 - \pi_i &= \Pr(Y_i = 0) \end{aligned} \tag{4.3}$$

the binary choice problem is restated as:

$$\begin{aligned} \pi_i &= \Pr(I_i^* > 0) \\ &= F(\varepsilon_i > -I_i) \\ &= 1 - F(-I_i) \end{aligned}$$

where $F(.)$ is a cumulative distribution function. Assuming the *pdf* associated with $F(.)$ is a symmetric distribution then:

$$\pi_i = F(I_i)$$

and hence:

$$1 - \pi_i = \Pr(I_i^* \leq 0) = 1 - F(I_i)$$

Specifying the distribution of ε_i and hence I_i^*, as normal, F is the cumulative normal function and the model is referred to as probit. For the case where ε_i is logistic, F is the cumulative logistic function and the model is the logit model.

For a sample of N individuals, the joint density function and hence the likelihood function, is given as:

$$L_F = \prod_{i=1}^{N} [F(I_i)]^{Y_i} [1 - F(I_i)]^{1-Y_i} \tag{4.4}$$

4.2 The Betit Model

Extending the specifications of the traditional probit and logit models can proceed on one of two fronts: by choosing specifications of π in (4.3) that give rise to more flexible response surfaces than the probit and logit specifications, or by adopting more flexible distributions of the error term ε in (4.1) than either the normal or logistic distributions. Observe that only one of these solutions is independent as one implies the other. However, noting this distinction helps to understand the motivation behind the recently proposed qualitative response models, including the betit model introduced in this chapter.

Previous extensions are based on adopting flexible distributions. Three broad classes of distributional models can be identified, ranging from parametric to nonparametric. Parametric approaches consist of choosing distributions that are indexed by a set of parameters which can exhibit a range of distributional shapes. Earlier examples of this approach are given by Prentice (1976) and Copenhaver and Mielke (1977). Semi-parametric approaches have been suggested by Cosslett (1983) and Klein and Spady (1987) where greater flexibility is achieved by replacing the parametric distribution by a kernel density. The semi-nonparametric approach of Gabler *et al.* (1993) falls between the parametric and semi-parametric approaches. It consists of using the flexible functional forms of Gallant and Nychka (1987) whereby the normal distribution is augmented by a Hermite polynomial of arbitrary order. A common feature of the above approaches is that a density is chosen, either parametric, or semi-parametric, or semi-nonparametric, and the response surface is specified as the corresponding cumulative distribution function.

In contrast to these previous extensions, this chapter follows Martin (1990) and approaches the problem of choosing a functional form of the response surface directly. From a practical point, approaching the problem directly yields functional forms which have the desired flexibility, but are less computationally demanding than many of the previously suggested extensions. To formalise the model, assume that the ith individual's subjective probability of choosing $Y = 1$ is considered to be a random variable p from a generalised beta distribution, as discussed in chapter 3, of the form:

$$f\left(p;\theta_{j,i}\right) = \exp\left[\theta_{1,i}\ln p + \theta_{2,i}\ln\left(1-p\right) + \theta_{3,i}p + \theta_{4,i}p^2 - \eta_i\right] \tag{4.5}$$

where $p \in (0,1)$, the $\theta_{j,i}$ terms are in general functions of the ith individual's attributes:

$$\theta_{j,i} = h_j(X_i\beta_j) \tag{4.6}$$

where $h_j(.)$ is a general function which can be linear or nonlinear and β_j has at most dimension K, the number of attributes. The normalising constant η_i is calculated as:

$$\eta_i = \ln\int_0^1 \exp\left[\theta_{1,i}\ln s + \theta_{2,i}\ln\left(1-s\right) + \theta_{3,i}s + \theta_{4,i}s^2\right] ds \tag{4.7}$$

to ensure that $\int f\left(p;\theta_{j,i}\right) dp = 1$.

The choice of a beta distribution is natural as p lies in the unit interval, whilst the generalised beta distribution provides the desired flexibility for estimating a richer class of response surfaces. The first two terms of the generalised beta distribution in (4.5), namely $\theta_{1,i}$ and $\theta_{2,i}$, represent the standard beta distribution. The remaining two terms, $\theta_{3,i}$ and $\theta_{4,i}$, distinguish the generalised beta distribution from the standard beta distribution. All moments exist provided that $\theta_{1,i}$, $\theta_{2,i} > -1$. As the generalized beta distribution is a subordinate of the generalised exponential family, it inherits all of the properties of this family; see Cobb *et al.* (1983) for an earlier treatment of the generalised beta distribution, as well as Lye and Martin (1993a) for a more recent discussion of the generalised exponential family. Higher order polynomials in (4.5) can also be considered for even greater flexibility. Some extensions of the generalised beta distribution are pursued in Section 4.4 in constructing normality tests.

For an individual drawing from this distribution it is natural to choose the mode of the distribution as this, in general, is the point associated with the region of highest probability.[2] For certain parameterisations (4.5) can exhibit more than one mode. The modes of (4.5) are derived in the following theorem.

Theorem 1 *The modes and antimodes of (4.5) are given as the roots of the cubic:*

$$M = \left\{ p \in (0,1) \left| \theta_1 + (\theta_3 - \theta_1 - \theta_2) p + (2\theta_4 - \theta_3) p^2 - 2\theta_4 p^3 = 0 \right. \right\} \quad (4.8)$$

Proof. The generalised beta distribution is a member of the exponential family which is written as:

$$f(p;\theta) = \exp\left[g(p;\theta) - \eta\right] \quad (4.9)$$

where $g(p;\theta)$ is an arbitrary additive function of arguments in p. As the first derivative of $f(p;\theta)$ with respect to p gives:

$$\frac{\partial f(p;\theta)}{\partial p} = \frac{\partial g(p;\theta)}{\partial p} f(p;\theta)$$

and $f(p;\theta)$ is strictly positive, the location of the modes is determined by the roots of $\partial g(p;\theta)/\partial p$. For the generalised beta distribution in (4.5):

$$g(p;\theta) = \theta_1 \ln(p) + \theta_2 \ln(1-p) + \theta_3 p + \theta_4 p^2 \qquad p \in (0,1)$$

and:

$$\frac{\partial g(p;\theta)}{\partial p} = \frac{\theta_1}{p} - \frac{\theta_2}{1-p} + \theta_3 + 2\theta_4 p$$

[2] Although the mode is advocated as the criterion to be used when constructing the response surface, an alternative is the mean of the distribution (4.5). However, a drawback of using the mean is that it requires numerical integration techniques which makes estimation computationally more involved. For this reason attention is focused on the mode of the generalised beta distribution.

where subscripts associated with the ith individual are removed for convenience. Setting this expression equal to zero and rearranging gives the required result.

For a sample of N individuals, the analogue of the likelihood function in (4.4) is:

$$L_M = \prod_{i=1}^{N} \left[M\left(X_i; \Psi\right)\right]^{Y_i} \left[1 - M\left(X_i; \Psi\right)\right]^{1-Y_i} \tag{4.10}$$

where M is defined by (4.8) as the mode from the generalised beta distribution in (4.5) and the set of parameters is given by: $\Psi = \{\theta_1, \theta_2, \theta_3, \theta_4\}$.

4.2.1 Parameterisation and Identification

By using the mode of the generalised beta distribution it is not possible to identify all of the parameters of the betit model without further restrictions. To see this, rewrite the cubic equation in (4.8) as:

$$p^3 - \left(1 - \frac{\theta_3}{2\theta_4}\right)p^2 - \left(\frac{\theta_3 - \theta_1 - \theta_2}{2\theta_4}\right)p - \frac{\theta_1}{2\theta_4} = 0 \tag{4.11}$$

assuming that $\theta_4 \neq 0$. As the cubic equations (4.11) and (4.8) have the same root(s), it is not possible to identify all four parameters $\theta_1, \theta_2, \theta_3, \theta_4$. Another way to view the problem is that there are at most three roots of a cubic to determine four parameters. A convenient parameterisation that achieves identification, as well as achieving some additional properties that are discussed below, is given by:

$$\begin{aligned} \theta_1 &= \theta_2 = \alpha_1 > 0 \\ \theta_3 &= -\alpha_2 + X\beta \\ \theta_4 &= \alpha_2 \end{aligned} \tag{4.12}$$

The restriction $\theta_1 = \theta_2 = \alpha_1 > 0$ prevents the generalised beta distribution from exhibiting poles at the end points and thus ensures numerical stability of the likelihood function (4.10). This restriction is also sufficient to achieve identification of all remaining parameters. The parameterisations of θ_3 and θ_4 are chosen to impose unimodality upon the generalised beta distribution which is required to ensure that the response surface is a continuous and monotonically increasing function of the index. The $-\alpha_2$ term in θ_3 centres the response surface such that when the index is zero a root at 0.5 occurs, which is consistent with logit and probit models. The appropriate parameter restrictions to ensure unimodality are given in the following theorem and corollary.

Theorem 2 *The generalised beta distribution (4.5) with the restriction* $\theta_1 = \theta_2 > 0$ *exhibits interior unimodality for the parameter space:*

$$\theta_4 \leq 4\theta_1 \tag{4.13}$$

Proof. As the distribution in (4.5) is of the exponential family form:

$$f(p;\theta) = \exp\left[g(p;\theta) - \eta\right] \qquad p \in (0,1)$$

where the exponential function constitutes a strictly monotonic transformation, attention is focused on:

$$g(p;\theta) = \theta_1 \ln p + \theta_2 \ln(1-p) + \theta_3 p + \theta_4 p^2$$

which is unimodal if it contains one turning point. The following partial derivatives will be needed:

$$\begin{aligned} g'(p;\theta) &= \frac{\theta_1}{p} - \frac{\theta_2}{1-p} + \theta_3 + 2\theta_4 p \\ g''(p;\theta) &= -\frac{\theta_1}{p^2} - \frac{\theta_2}{(1-p)^2} + 2\theta_4 \end{aligned}$$

If $f(p;\theta)$ is unimodal then $g(.)$ contains one turning point, the global maximum, which implies $g'(.)$ has one real root in the unit interval and that $g''(.)$ is strictly negative at this point. This implies that the absence of real roots within the unit interval from $g''(.)$ constitutes a necessary and sufficient condition for unimodality. Setting the above expression of $g''(.)$ equal to zero and rearranging yields:

$$p^4 - 2p^3 - \left(\frac{\theta_1+\theta_2}{2\theta_4} - 1\right)p^2 + \frac{\theta_1}{\theta_4}p - \frac{\theta_1}{2\theta_4} = 0 \tag{4.14}$$

Assuming $\theta_2 = \theta_1$ enables (4.14) to be simplified to

$$p^4 - 2p^3 - \left(\frac{\theta_1}{\theta_4} - 1\right)p^2 + \frac{\theta_1}{\theta_4}p - \frac{\theta_1}{2\theta_4} = 0$$

which is written more conveniently as:

$$p^4 - 2p^3 - (k-1)p^2 + kp - 0.5k = 0 \tag{4.15}$$

where $k = \theta_1/\theta_4$. Equation (4.15) is factored as:

$$\left(p^2 - p + \lambda_1\right)\left(p^2 - p + \lambda_2\right) = 0 \tag{4.16}$$

where:

$$\lambda_1, \lambda_2 = \frac{-k \pm (k^2 + 2k)^{0.5}}{2}$$

Therefore, the four roots of (4.16) are summarised as:

$$p = 0.5 \pm 0.5 (1 - 4\lambda_1, \lambda_2)^{0.5} \tag{4.17}$$

If $k > 0.25$ then two roots are complex and two are real, though outside the unit interval. If $k = 0.25$ there are two repeated roots at $p = 0.5$ and two real roots outside the unit interval. If $0 < k < 0.25$ there are two real roots within the unit interval and two outside, that is the generalised beta distribution is bimodal. If $k = 0$ then there are two repeated roots at $p = 0$ and $p = 1$. If $k < 0$ then all roots are complex. Adopting $k > 0.25$, and given that $\theta_1 = \theta_2 > 0$ is imposed to ensure poles do not occur at boundaries, the necessary and sufficient conditions for unimodality of the generalised beta distribution is given by (4.13).

Corollary 3 *Given the parameterisations in (4.12), the condition for unimodality reduces to:*

$$\alpha_2 \leq 4\alpha_1 \tag{4.18}$$

Proof. The result immediately follows from substituting the relevant terms in (4.12) into (4.13).

The response surfaces of the probit and logit models are compared in Figure 4.1 with the betit model for the parameterisations in (4.12) where $\alpha_1 = 0.25$ and $\alpha_2 = 1$. All response surfaces are symmetrical around $I = 0$, where each yield probabilities of $\pi = 0.5$. The main difference between these models is that the betit response surface approaches the upper and lower bounds, that is $\pi = (0, 1)$, more slowly than either the probit or logit specifications, whereas in the neighbourhood of $I = 0$, the slope of the betit response surface is steeper than the corresponding response surfaces of either of the probit and logit models.

The results in Figure 4.1 suggest that if the true process is represented by the betit response surface and a cumulative distribution is to be used for modelling the response surface, then the appropriate form of the probability distribution of ε in (4.1) is leptokurtic. This result follows from the observation that a steeper response surface around zero is associated with a distribution with a sharper peak than the normal distribution, whereas the slowness of the response surface to approach the upper and lower boundaries is representative of fat tails in the distribution of ε relative to the normal distribution.

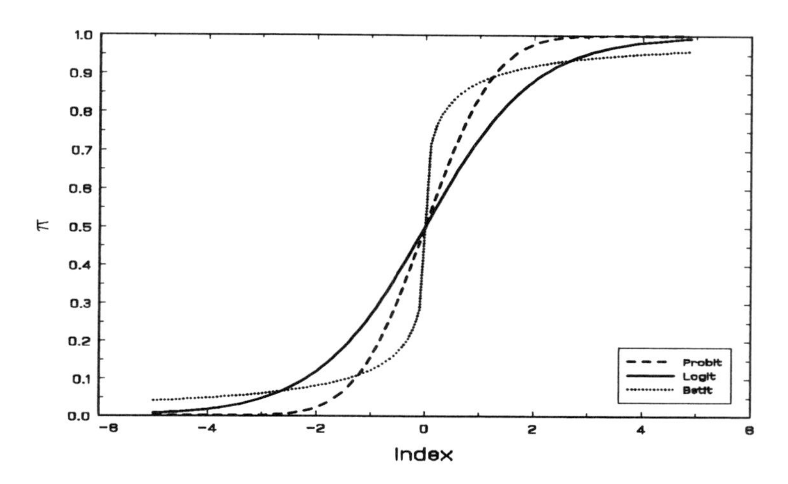

Figure 4.1: Probit, logit and betit response surfaces

4.2.2 Relationships with Existing Approaches

Probit

Consider the restriction $\theta_1 = \theta_2 = 0$, in (4.5) which results in the generalised beta distribution being bell-shaped as is typical of the normal density:

$$f(p, \theta) = \exp\left[\theta_3 p + \theta_4 p^2 - \eta\right] \qquad p \in (0, 1) \tag{4.19}$$

This distribution has a single mode which is given by:

$$M = -\frac{\theta_3}{2\theta_4} \tag{4.20}$$

A parameterisation that is motivated by the form of the standard normal distribution is to set $\theta_4 = -0.5$, thereby reducing the mode to:

$$M = \theta_3 \tag{4.21}$$

To ensure that the mode falls within the unit interval, an appropriate transformation is to restrict θ_3 by using a cumulative density function. Letting $\theta_3 = X\beta$ and transforming with a cumulative density function gives the mode as:

$$M = \theta_3 = F(X\beta) \tag{4.22}$$

By setting F equal to the cumulative normal distribution yields the probit model.

Logit

An obvious method of embedding the logit model in the betit framework is to specify F in (4.22) as the cumulative logistic function. An alternative approach for embedding the logit model is to begin with the standard beta density by setting $\theta_3 = \theta_4 = 0$ in (4.5) to give:

$$f(p, \theta) = \exp\left[\theta_1 \ln p + \theta_2 \ln(1-p) - \eta\right], \qquad p \in (0,1) \tag{4.23}$$

From (4.8) this yields the single mode:

$$M = \frac{\theta_1}{\theta_1 + \theta_2} \tag{4.24}$$

By adopting the parameterisation:

$$\begin{aligned} \theta_1 &= 1 \\ \theta_2 &= \exp\left[-X\beta\right] \end{aligned}$$

in (4.24) yields the mode:

$$M = \frac{1}{1 + \exp\left[-X\beta\right]}$$

This expression is the cumulative logistic distribution which represents the logit model. The relationship between the logit and the standard beta density has also been noted by Heckman and Willis (1977, p.41).

Heterogeneity

The specification of a distribution for p in (4.5) is akin to the models of heterogeneity introduced by Heckman and Willis (1977) whereby the probability of selecting an event is a function of unobservable variables specific to the ith individual. Although models of heterogeneity have primarily been developed for panel data, the relationship with the betit model still holds for micro data based on one time period.

To show this, consider the model of Heckman and Willis (1977) where the probability that $Y_i = 1$ is assumed to be conditioned on a set of factors X_i:

$$P(Y_i = 1 | X_i) = p_i \tag{4.25}$$

To ensure that the conditional probabilities in (4.25) are in the unit interval, p_i is chosen to have the beta distribution

$$f(p;\theta_{j,i}) = \exp\left[\theta_{1,i}\ln p + \theta_{2,i}\ln(1-p) - \eta_i\right] \tag{4.26}$$

which is a special case of (4.5) where $\theta_{3,i} = \theta_{4,i} = 0$. In contrast to the derivation of the betit model where the mode of the distribution is chosen to construct the response surface, Heckman and Willis choose the mean

$$\Xi_i = \frac{1+\theta_{1,i}}{2+\theta_{1,i}+\theta_{2,i}} \tag{4.27}$$

as $E\left[Y_i = 1\right] = \Xi_i$. Now choosing the parameterisation:

$$\begin{aligned} \theta_{1,i} &= \exp\left[-X_i\alpha_1\right] - 1 \\ \theta_{2,i} &= \exp\left[-X_i\alpha_2\right] - 1 \end{aligned}$$

in (4.27) gives the logit model:

$$\Xi_i = \frac{1}{1+\exp\left[-X\beta\right]}$$

where $\beta = \alpha_2 - \alpha_1$.

Semi-nonparametrics

The adoption of a flexible parametric density for ε in (4.1) is related to the nonparametric class of estimators considered by Cosslett (1983), Klein and Spady (1987), and Gabler *et al.* (1993); see also Pagan and Ullah (1996) for a review of this literature. In the case of the semi-nonparametric approach of Gabler *et al.* a cumulative density function of a semi-nonparametric probability density is used to define the binary choice response surface. This differs from the betit approach which defines the binary choice response surface as the mode of a generalised beta distribution. However, there are distributional similarities between these approaches.

The density of ε in equation (4.1) adopted by Gabler *et al.* (1993) is a normal distribution augmented with a Hermite polynomial of the form:

$$f_{SNP}(\varepsilon;\alpha) = \hat{\eta}\left((2\pi)^{-1/4} + \sum_{i=1}^{K}\alpha_i\varepsilon^i\right)^2 \exp\left[-0.5\varepsilon^2\right] \tag{4.28}$$

where $\hat{\eta}$ is the normalising constant and the Kth order polynomial is squared to ensure positivity. The α parameters correspond to the Hermite polynomial terms which control the distributional characteristics. Setting $\alpha_i = 0$, $\forall i$, in (4.28) yields the normal distribution.

The generalised beta distribution can also be expressed as a product of a 'normal' and a polynomial. Rewriting (4.5) as:

$$f(p;\theta) = \hat{\eta} \exp\left[\theta_1 \ln(p) + \theta_2 \ln(1-p) + \theta_3 p + (\theta_4 + 0.5) p^2\right] \exp\left[-0.5p^2\right]$$

and applying a Taylor series expansion to the square root of the first exponential expression (around the mode of the distribution), and squaring the approximating polynomial yields:

$$f(p;\gamma) \simeq \hat{\eta} \left(\gamma_0 + \sum_{i=1}^{K} \gamma_i p^i\right)^2 \exp\left[-0.5p^2\right] \tag{4.29}$$

where the γ_i, $i = 0, 1, ..., K$, parameters are functions of θ. The parameter γ_0 is the normalising constant which ensures that the density integrates to unity. A comparison of (5.9) with (4.28) shows that the γ parameters in (5.9) and the α parameters in (4.28) operate in a similar way; namely, both parameter sets control the shape of the distribution.

The betit modelling framework has a computational advantage over the semi-nonparametric approach. In contrast to the latter approach, estimation of the betit response surface does not require numerical integration routines to compute the normalising constant $\hat{\eta}_i$ in (4.28) $\forall i$. For relatively large micro data sets these computations can become demanding.

4.3 Pseudo Maximum Likelihood Estimation

For a sample of N individuals, the betit log-likelihood function is given by taking the logarithms of (4.10) to yield:

$$\ln L = \sum_{i=1}^{N} Y_i \ln\left[M(X_i;\Psi)\right] + \sum_{i=1}^{N} (1 - Y_i) \ln\left[1 - M(X_i;\Psi)\right] \tag{4.30}$$

where $\Psi = \{\beta_1, \beta_2, \beta_3, \beta_4\}$ is the set of unknown parameters and M is the mode of the generalised beta distribution. The pseudo maximum likelihood parameter estimates of Ψ are obtained by solving the set of partial derivatives:

$$\frac{\partial \ln L}{\partial \Psi} = \sum_{i=1}^{N} \frac{Y_i - M_i}{M_i(1 - M_i)} \frac{\partial M_i}{\partial \Psi} = 0 \tag{4.31}$$

An analytical expression for $\partial M_i/\partial\Psi$ is obtained by applying the implicit function theorem to the cubic polynomial (4.8), denoted $P(\theta, \Psi)$, to yield:

$$\begin{aligned} \frac{\partial M_i}{\partial \Psi} &= -\frac{\partial P/\partial \Psi}{\partial P/\partial M_i} \\ &= -\left(\frac{\partial P}{\partial \theta}\frac{\partial \theta}{\partial \Psi}\right) \Big/ \frac{\partial P}{\partial M_i} \end{aligned} \tag{4.32}$$

Noting that $\Psi = \{\beta_1, \beta_2, \beta_3, \beta_4\}$, equation (4.32) can be written in matrix form as:

$$\frac{\partial M_i}{\partial \Psi} = -\left(\frac{\partial P}{\partial M_i}\right)^{-1} \left[\begin{array}{cccc} \frac{\partial P}{\partial \theta_1} & \frac{\partial P}{\partial \theta_2} & \cdots & \frac{\partial P}{\partial \theta_4} \end{array}\right] \left[\begin{array}{ccc} \frac{\partial \theta_1}{\partial \beta_{1,1}} & \cdots & \frac{\partial \theta_1}{\partial \beta_{1,K}} \\ \vdots & \ddots & \vdots \\ \frac{\partial \theta_4}{\partial \beta_{4,1}} & \cdots & \frac{\partial \theta_4}{\partial \beta_{4,K}} \end{array}\right] \tag{4.33}$$

where:

$$\frac{\partial P}{\partial \theta_1} = 1 - M_i \qquad \frac{\partial P}{\partial \theta_3} = M_i - M_i^2$$

$$\frac{\partial P}{\partial \theta_2} = -M_i \qquad \frac{\partial P}{\partial \theta_4} = 2\left(M_i^2 - M_i^3\right)$$

and:

$$\frac{\partial P}{\partial M_i} = \theta_3 - \theta_1 - \theta_2 + 2\left(2\theta_4 - \theta_3\right) M_i - 6\theta_4 M_i^2$$

$\partial\theta/\partial\Psi$ depends on the exogenous variable specification adopted.

Additional structure can be applied to the above expression by applying parameterisations (4.12) and (4.18). The no-poles restriction $\theta_1 = \theta_2 = \alpha_1 > 0$ and the inequality restriction needed for unimodality, $\alpha_2 \leq 4\alpha_1$, are achieved by the parameterisation:

$$\alpha_1 = 0.25, \ \alpha_2 = F(\delta) \tag{4.34}$$

where F is the cumulative normal distribution function and δ is a parameter to be estimated. The choice of $\alpha_1 = 0.25$ is arbitrary as the key condition is that $\alpha_2 \leq 4\alpha_1$. Combining equation (4.34) with (4.12) redefines cubic equation (4.8) to give the mode, M, as the value which satisfies:

$$\frac{1}{8F(\delta)} + \left(\frac{X\beta - 0.5}{2F(\delta)} - 0.5\right) M + \left(1.5 - \frac{X\beta}{2F(\delta)}\right) M^2 - M^3 = 0 \tag{4.35}$$

Therefore, $\partial P/\partial\theta$ expressions remain unchanged, however:

$$\frac{\partial P}{\partial M_i} = -0.5 + \frac{X\beta - 0.5}{2F(\delta)} + \left(3 - \frac{X\beta}{F(\delta)}\right) M - 3M^2$$

and:

$$\begin{bmatrix} \frac{\partial\theta_1}{\partial\beta_{1,1}} & \cdots & \cdots & \frac{\partial\theta_1}{\partial\beta_{1,K}} \\ \frac{\partial\theta_2}{\partial\beta_{2,1}} & \ddots & & \frac{\partial\theta_2}{\partial\beta_{2,K}} \\ \frac{\partial\theta_3}{\partial\beta_{3,1}} & & \ddots & \frac{\partial\theta_3}{\partial\beta_{3,K}} \\ \frac{\partial\theta_4}{\partial\beta_{4,1}} & \cdots & \cdots & \frac{\partial\theta_4}{\partial\beta_{4,K}} \end{bmatrix} = \begin{bmatrix} 0 & \cdots & \cdots & \cdots & 0 \\ 0 & \cdots & \cdots & \cdots & 0 \\ -f(\delta) & 1 & X_1 & \cdots & X_K \\ f(\delta) & 0 & \cdots & \cdots & 0 \end{bmatrix}$$

The asymptotic properties of the pseudo maximum likelihood estimator are derived from the theory of misspecified models as developed by White (1982). For the set of parameters $\Psi = \{\beta_1, \beta_2, \beta_3, \beta_4\}$, the pseudo true value Ψ^* is defined as:

$$\Psi^* = \arg\max_{\Psi} \underset{X}{E}\,\underset{0}{E} \ln L$$

where $\ln L$ is defined by (4.30) and the second expectation is taken with respect to the true model with parameter vector Ψ^0. Under standard regularity conditions, the maximum likelihood estimator $\widehat{\Psi}$ converges to the pseudo true value Ψ^*, and the asymptotic distribution is:

$$\sqrt{N}\left(\widehat{\Psi} - \Psi^*\right) \xrightarrow{d} N\left(0,\ I(\Psi^*)^{-1} G(\Psi^*) I(\Psi^*)^{-1}\right) \qquad (4.36)$$

where $I(\Psi^*)$ is the information matrix and $G(\Psi^*)$ is the outer product of the gradients of the log-likelihood.[3] Consistent estimates of these expressions are obtained by evaluating I and G at $\widehat{\Psi}$.

4.4 Constructing Normality Tests

As was shown in subsection 4.2.2, the betit model can nest the probit model by an appropriate parameterisation of (4.5). Given that (4.5) can be extended to allow for a richer set of parameterisations, this suggests that there are several ways to nest the normal distribution in the betit model. No matter which parameterisation is adopted, the Lagrange multiplier framework in conjunction with the betit model provides a convenient method for contructing alternative tests of normality and hence the suitability of the probit model, in the spirit of the Bera *et al.* (1984) normality test; see also chapter 3 for a discussion of testing for normality in truncated distributional models.

[3]If the generalised beta distribution in (4.5) does not encompass the true distribution, the maximum likelihood estimator is asymptotically biased. Allowing the order of the generalised beta distribution to increase with the sample size enables the degree of asymptotic bias to decrease.

The LM test statistic is defined as:

$$LM = \left(\frac{\partial \ln L}{\partial \Psi}\right) E\left[\frac{\partial \ln L}{\partial \Psi}\frac{\partial \ln L}{\partial \Psi'}\right]^{-1} \left(\frac{\partial \ln L}{\partial \Psi'}\right)\Bigg|_{\Psi=\Psi_0} \qquad (4.37)$$

where the expectations operator $E[.]$ and all derivatives are evaluated under the null hypothesis H_0, that $\Psi = \Psi_0$. This statistic is asymptotically distributed as χ_R^2 under the null hypothesis, where R is the number of restrictions; see Bera *et al.* (1984) for a discussion of the necessary regularity conditions for this result to hold.

In the case of the betit model considered so far, from the discussion in subsection 4.2.2 an appropriate parameterisation of the betit model in (4.5) for constructing a test of the probit model is:

$$f(p;\theta_{j,i}) = \exp\left[\theta_{1,i}\ln p + \theta_{2,i}\ln(1-p) + F(X_i\beta)\,p - 0.5p^2 - \eta_i\right] \qquad (4.38)$$

where $F(X\beta)$ is the cumulative normal distribution. The parameter set is $\Psi = (\theta_1, \theta_2, F(X\beta), -0.5)$,and the restricted parameter set is $\Psi_0 = (0, 0, F(X\beta), -0.5)$. The first order derivatives of the log-likelihood function (4.30) are derived from equation (4.11) by application of the implicit function theorem, which is used to derive $\partial M/\partial\Psi$ terms in equation (4.31), as:

$$\begin{aligned}
\frac{\partial \ln L}{\partial \beta} &= \sum_{i=1}^{N} \Lambda_i f_i X_i \\
\frac{\partial \ln L}{\partial \theta_1} &= \sum_{i=1}^{N} \frac{\Lambda_i}{M_i} \\
\frac{\partial \ln L}{\partial \theta_2} &= \sum_{i=1}^{N} \frac{\Lambda_i}{M_i - 1}
\end{aligned} \qquad (4.39)$$

where $f(.)$ denotes the normal probability density function associated with $F(.)$, X is of dimension $N \times K$ and:

$$\Lambda_i = \frac{M_i - Y_i}{F_i - \theta_1 - \theta_2 - 2(1+F_i)M_i + 3M_i^2}$$

Under the null hypothesis $M_i = F_i$, the cumulative normal distribution, and the first order derivatives reduce to:

$$\frac{\partial \ln L}{\partial \beta}\bigg|_{\Psi=\Psi_0} = 0$$

$$\begin{aligned}\left.\frac{\partial \ln L}{\partial \theta_1}\right|_{\Psi=\Psi_0} &= \sum_{i=1}^{N} \frac{F_i - Y_i}{F_i^2 (F_i - 1)} \\ \left.\frac{\partial \ln L}{\partial \theta_2}\right|_{\Psi=\Psi_0} &= \sum_{i=1}^{N} \frac{F_i - Y_i}{F_i (F_i - 1)^2}\end{aligned} \tag{4.40}$$

The information matrix under the null hypothesis is derived from the first order derivatives in (4.39) and contains the terms:

$$\begin{aligned}E\left[\frac{\partial \ln L}{\partial \beta}\frac{\partial \ln L}{\partial \beta'}\right]\Big|_{\Psi=\Psi_0} &= \sum_{i=1}^{N} \frac{f_i{}^2 X_i X_i'}{F_i (F_i - 1)} \\ E\left[\frac{\partial \ln L}{\partial \theta_1}\frac{\partial \ln L}{\partial \beta'}\right]\Big|_{\Psi=\Psi_0} &= \sum_{i=1}^{N} \frac{f_i X_i}{F_i^2 (F_i - 1)} \\ E\left[\frac{\partial \ln L}{\partial \theta_2}\frac{\partial \ln L}{\partial \beta'}\right]\Big|_{\Psi=\Psi_0} &= \sum_{i=1}^{N} \frac{f_i X_i}{F_i (F_i - 1)^2} \\ E\left[\frac{\partial \ln L}{\partial \theta_1}\frac{\partial \ln L}{\partial \theta_1'}\right]\Big|_{\Psi=\Psi_0} &= \sum_{i=1}^{N} \frac{1}{F_i^3 (F_i - 1)} \\ E\left[\frac{\partial \ln L}{\partial \theta_2}\frac{\partial \ln L}{\partial \theta_2'}\right]\Big|_{\Psi=\Psi_0} &= \sum_{i=1}^{N} \frac{1}{F_i (F_i - 1)^3} \\ E\left[\frac{\partial \ln L}{\partial \theta_1}\frac{\partial \ln L}{\partial \theta_2'}\right]\Big|_{\Psi=\Psi_0} &= \sum_{i=1}^{N} \frac{1}{F_i^2 (F_i - 1)^2}\end{aligned} \tag{4.41}$$

which is based on the result that: $E\left[(Y_i - E[Y_i])^2\right]\big|_{\Psi=\Psi_0} = F_i (1 - F_i)$.

Substituting the first order derivatives in (4.40) and the second order derivatives in (4.41) into (4.37) gives the test statistic. This test is identified as $GB-1$ in recognition that it is based on the generalised beta distribution. Inspection of the expressions making up the $GB-1$ test show that it can be easily computed in standard regression packages as it just requires the computation of F_i, the cumulative normal distribution. As the null hypothesis contains two restrictions the LM test statistic is asymptotically distributed as chi-squared with two degrees of freedom. Large values of the test statistic constitute rejection of the null hypothesis of normality.

The Bera *et al.* (1984) test of normality constitutes a joint test of the third and fourth moments of the distribution. Following the analysis of Lye and Martin (1994) who showed the relationship between the Jarque and Bera test of normality and generalised exponential distributions, an alternative specification of the generalised beta distribution given in (4.5) is to replace the $\ln(p)$ and $\ln(1-p)$ terms by p^4 and p^5, thereby yielding:

$$f(p;\theta_{j,i}) = \exp\left[\theta_{3,i}p + \theta_{4,i}p^2 + \theta_{6,i}p^4 + \theta_{7,i}p^5 - \eta_i\right]; \quad p \in (0,1) \qquad (4.42)$$

with $\theta_3 = F(X\beta)$ and $\theta_4 = -0.5$. As is shown below, when the mode of the density is used to construct the response surface, the inclusion of the fourth and fifth order terms in (4.42) yields a Lagrange multiplier test of the third and fourth moments similar to the Bera *et al.* form. Alternatively, if the mean of the distribution is used in constructing the response surface it is appropriate to use the third and fourth power terms in (4.42).

Under the null hypothesis $(\theta_6, \theta_7) = 0$, so the restricted parameter set is $(\theta_3, \theta_4, \theta_6, \theta_7) = (F(X\beta), -0.5, 0, 0)$. Given that the mode of the density in (4.42) is defined as the root of:

$$\theta_{3,i} + 2\theta_{4,i}M_i + 4\theta_{6,i}M_i^3 + 5\theta_{7,i}M_i^4 = 0$$

the implicit function theorem is used to derive $\partial M/\partial\Psi$ and hence from (4.31) the first order conditions are:

$$\begin{aligned}
\frac{\partial \ln L}{\partial \beta} &= \sum_{i=1}^{N} \frac{Y_i - M_i}{M_i(1-M_i)} f_i X_i \\
\frac{\partial \ln L}{\partial \theta_6} &= \sum_{i=1}^{N} \frac{Y_i - M_i}{M_i(1-M_i)} 4M_i^3 \\
\frac{\partial \ln L}{\partial \theta_7} &= \sum_{i=1}^{N} \frac{Y_i - M_i}{M_i(1-M_i)} 5M_i^4 \qquad (4.43)
\end{aligned}$$

Under the null hypothesis these expressions reduce to:

$$\begin{aligned}
\left.\frac{\partial \ln L}{\partial \beta}\right|_{\Psi=\Psi_0} &= 0 \\
\left.\frac{\partial \ln L}{\partial \theta_6}\right|_{\Psi=\Psi_0} &= \sum_{i=1}^{N} \frac{Y_i - F_i}{F_i(1-F_i)} 4F_i^3 \\
\left.\frac{\partial \ln L}{\partial \theta_7}\right|_{\Psi=\Psi_0} &= \sum_{i=1}^{N} \frac{Y_i - F_i}{F_i(1-F_i)} 5F_i^4 \qquad (4.44)
\end{aligned}$$

Proceeding as with the derivation of the previous test statistic, the information matrix under the null hypothesis now contains the terms:

$$E\left[\frac{\partial \ln L}{\partial \beta}\frac{\partial \ln L}{\partial \beta'}\right]\Bigg|_{\Psi=\Psi_0} = \sum_{i=1}^{N} \frac{f_i{}^2 X_i X_i'}{F_i(F_i - 1)}$$

$$E\left[\frac{\partial \ln L}{\partial \theta_6}\frac{\partial \ln L}{\partial \beta'}\right]\Bigg|_{\Psi=\Psi_0} = \sum_{i=1}^{N}\frac{4F_i^3 f_i X_i}{F_i\left(F_i-1\right)}$$
$$E\left[\frac{\partial \ln L}{\partial \theta_7}\frac{\partial \ln L}{\partial \beta'}\right]\Bigg|_{\Psi=\Psi_0} = \sum_{i=1}^{N}\frac{5F_i^4 f_i X_i}{F_i\left(F_i-1\right)}$$
$$E\left[\frac{\partial \ln L}{\partial \theta_6}\frac{\partial \ln L}{\partial \theta_6'}\right]\Bigg|_{\Psi=\Psi_0} = \sum_{i=1}^{N}\frac{16F_i^6}{F_i\left(F_i-1\right)}$$
$$E\left[\frac{\partial \ln L}{\partial \theta_7}\frac{\partial \ln L}{\partial \theta_7'}\right]\Bigg|_{\Psi=\Psi_0} = \sum_{i=1}^{N}\frac{25F_i^8}{F_i\left(F_i-1\right)}$$
$$E\left[\frac{\partial \ln L}{\partial \theta_6}\frac{\partial \ln L}{\partial \theta_7'}\right]\Bigg|_{\Psi=\Psi_0} = \sum_{i=1}^{N}\frac{20F_i^7}{F_i\left(F_i-1\right)} \tag{4.45}$$

Substituting the first order derivatives in (4.44) and the second order derivatives in (4.45) into (4.37) gives the second form of the test statistic. The second test of normality is identified as $GB-2$, which is asymptotically distributed as χ_2^2 under the null hypothesis. Given that this test is explicitly based on testing the third and fourth order moments, it is anticipated that it should yield similar qualitative results to the Bera *et al.* normality test. Inspection of the first order conditions in (4.44) shows that the test statistic has a RESET interpretation as it involves computing the covariance between the residual $Y_i - F_i$, and the third and fourth order powers of F_i, with all terms weighted by $\left(F_i\left(1-F_i\right)\right)^{-1/2}$. This is similar to the normality test suggested by Ruud (1983) except that the terms F_i^3 and F_i^4 are replaced by $f_i I_i^3$ and $f_i I_i^4$, where f_i is the normal density and $I_i = X_i\beta$ is the index; see also Pagan and Vella (1989).

4.5 Female Labour Force Participation

The performance of the betit model is now compared with the probit and logit models in the context of married female labour force participation. The number of obervations is 753 which are taken from Berndt (1991, p.651), which in turn represents a sample from the 1976 Panel Study of Income Dynamics. The dependent variable is $Y=1$ if the wife is participating in the labour force and $Y=0$ otherwise. Of the 753 observations, 428 women are recorded as participants, $Y=1$, and 325 as non-participants, $Y=0$. Following the specification adopted by Berndt (1991, p.655), labour force participation is modelled as a function of a constant (C), the log of wife's average hourly earnings ($LWW1$), the number of children less than 6 years old in the household ($KL6$), the number of children between ages 6 and 18 in

Table 4.1: Estimates of the female labour force model[(a)]

	Logit	Probit	Betit
C	0.9509	0.5679	0.0998
	(0.8036)	(0.4794)	(0.0160)
$LWW1$	0.4644	0.2820	0.0706
	(0.1691)	(0.0995)	(0.0071)
$KL6$	−1.4692	−0.8808	−0.2544
	(0.2062)	(0.1174)	(0.0254)
$K618$	−0.0512	−0.0297	−0.0112
	(0.0716)	(0.0422)	(0.0012)
WA	−0.0583	−0.0350	−0.0104
	(0.0129)	(0.0076)	(0.0010)
WE	0.2120	0.1277	0.0396
	(0.0459)	(0.0267)	(0.0040)
UN	−0.0186	−0.0111	−0.0015
	(0.0267)	(0.0160)	(0.0002)
CIT	0.0128	0.0100	−0.0093
	(0.1831)	0.1046)	(0.0017)
$PRIN$	−0.0353	−0.0212	−0.0053
	(0.0088)	(0.0051)	(0.0006)
δ			5.4400

(a) Heteroscedastic-consistent standard errors are in parentheses. The standard error for δ is not computed to improve the accuracy of the remaining standard error estimates.

the household ($K618$), wife's age (WA), wife's educational attainment (WE), the percentage point unemployment rate in country of residence (UN), a dummy variable which equals 1 if living in a large city, otherwise 0 (CIT), and wife's property income scaled by 1000 ($PRIN$).

Parameter estimates are computed in GAUSS VMi version 3.2.13 with the BFGS algorithm from the MAXLIK library version 4.0.16. Standard errors are based on (4.36). The mode of the betit model is computed by (4.35).

The parameter estimates and standard errors of the logit, probit and betit models are presented in Table 4.1. The diagnostics are given in Table 4.2. Of the three tests of normality given in Table 4.2, the Bera *et al.* (1984) test of normality (BJL) and $GB-2$ tests yield similar qualitative conclusions,

Table 4.2: Diagnostics of the female labour force model

	Logit	Probit	Betit
$\ln L$	-449.4757	-449.3970	-444.8687
$\ln L$ (zero slopes)	-514.8732	-514.8732	-514.8732
sum sqd. resid.	154.9090	155.0118	151.5562
Efron's R^2	0.1614	0.1609	0.1796
Lave's R^2	0.1513	0.1507	0.1696
McFadden's R^2	0.1270	0.1272	0.1360
Akaike I.C.	458.4757	458.3970	453.8687
BJL		8.6953	
GB-1 (based on (4.38))		1.0343	
GB-2 (based on (4.42))		11.1172	

Table 4.3: Prediction table of the female labour force model

		Predicted						
		$Y=0$			$Y=1$			
		Logit	Probit	Betit	Logit	Probit	Betit	Total
Actual	$Y=0$	164	160	212	161	165	113	325
	$Y=1$	82	79	98	346	349	330	428
Total		246	239	310	507	514	443	753

namely the rejection of the null hypothesis of normality at the 5 per cent nominal level of significance. In contrast, the $GB-2$ test fails to reject the null. This result serves to reinforce the relationship between the BJL and $GB-2$ tests. It also may suggest that the $GB-1$ test lacks power when testing for normality.[4]

The goodness-of-fit statistics reported in Table 4.2 show that the betit model gives rise to a small improvement for within sample prediction over the probit and logit specifications. These gains are highlighted in Table 4.3 which demonstrates that the logit and probit models yield similar predictions, correctly predicting 510 and 509 respectively of 753 observations, compared to 542 correct predictions made by the betit model. These results are supported by the model specification in Table 4.2 which shows that the betit model minimises the Akaike information criterion.

[4]Juxtaposed with this result is the result that if the information matrix is approximated by the outer product of the likelihoods across the obervations, the computed value of the $GB-1$ test statistic is 13.4417, which agrees with the qualitative results of the BJL and the $GB-2$ tests.

4.6 Conclusions

This chapter introduced a new class of models for binary data called betit. The betit framework was demonstrated to have many advantages over the traditional probit and logit binary models, as well as some of the recent extensions suggested in the literature. The superiority of the betit model over the probit and logit specifications was demonstrated in an empirical application of married female labour force participation.

Chapter 5

Estimation of Generalised Distributions

Alex Bakker

Estimators of the generalised exponential class of distributions have focused on the situation where data are ungrouped. In the case where data are grouped in the form of frequencies, Martin (1990) proposed the least squares frequency estimator (LSFE). Given that estimates are obtained using standard linear regression analysis, the approach is straightforward and overcomes the computational difficulties associated with maximum likelihood estimation. However, properties of the LSFE, as stated in Martin (1990), are in general unknown.

This chapter provides a more formal derivation of the LSFE. Additional contributions include formal derivation of residual properties, the LSFE covariance matrix and conditions required for unbiasedness and consistency. Using the covariance matrix an efficient generalised least squares frequency estimator (GLSFE) is proposed. The small sample properties of the estimators and standard errors are examined via Monte Carlo experiments applied to the generalised gamma distribution. In addition, the optimal relationship between sample size and the number of class intervals for both estimators is obtained by deriving a numerical approximation to the mean squared error criterion.

Section 5.1 derives the LSFE. The error properties of the LSFE are outlined in Section 5.2. Section 5.3 examines the econometric properties of the LSFE and introduces the GLSFE. The optimal relationship between sample size and number of class intervals applied when calculating frequencies is discussed with reference to the LSFE and GLSFE in Section 5.4. Section 5.5 provides an examination of the sampling distribution of the standard

t-statistics associated with the parameter estimates of the frequency estimators.

5.1 Derivation of the LSFE

The least squares frequency estimator (LSFE) is used to obtain parameter estimates of continuous generalised exponential family unconditional densities. The generalised exponential family of density functions, $f(y)$, for some random variable y, is defined as:

$$f(y) = \exp\left[\sum_{j=1}^{J} \theta_j \psi_j(y) - \eta\right] \qquad y \in D \tag{5.1}$$

where $\psi_j(.)$ is some function depending on the density type, θ_j are the associated parameter coefficients, J is the number of additive terms in the density and η is the normalising constant given by:

$$\eta = \ln \int_D \exp\left[\sum_{j=1}^{J} \theta_j \psi_j(y)\right] dy \tag{5.2}$$

thus ensuring that $\int_D f(y)\, dy = 1$.

A distinguishing feature of the LSFE is that it is based on frequencies. This contrasts with the least squares estimator of the Pareto distribution suggested by Johnson and Kotz (1970, p.235), which is based on cumulative frequencies; see Finch (1989), Martin (1990) and Lye and Martin (1994). The error terms resulting from the LSFE are more likely to satisfy the assumption of independence, a crucial assumption for consistency of the least squares estimator.

To use equation (5.1) in a frequency context it is necessary to divide the density's domain, D, into K class intervals, such that the probability associated with the kth class interval is:

$$p_k = \int_{a_k}^{a_k+W_k} \exp\left[\sum_{j=1}^{J} \theta_j \psi_j(y) - \eta\right] dy \tag{5.3}$$

where a_k denotes the lower bound of the kth class interval of width W_k. For a sample of size N, the natural logarithm of the expected frequency in the kth class interval is:

$$\ln E_k = \ln(Np_k) \tag{5.4}$$

The LSFE is derived from two approximations of $\ln E_k$.

The first approximation of $\ln E_k$ arises from computing the integral in equation (5.3) numerically. This expression can be conveniently written as:

$$\ln E_k = \ln A_k + \varepsilon_{1,k} \tag{5.5}$$

where $\varepsilon_{1,k}$ is the numerical integration or 'quadrature' error. The term A_k represents a rectangular area approximation of the integral in equation (5.3) and is calculated as:

$$A_k = NW_k f(y_k) \tag{5.6}$$

where $y_k = a_k + 0.5W_k$ is the midpoint of the kth class.

The second approximation of $\ln E_k$ represents an empirical approximation and is written:

$$\ln E_k = \ln O_k + \varepsilon_{2,k} \tag{5.7}$$

where O_k is the empirical frequency observed for the kth class interval and $\varepsilon_{2,k}$ is the empirical sampling error of the kth class interval. It is assumed that the sample size is sufficiently large such that:

$$\int_{\min D}^{\min \tilde{y}} f(y)\, dy + \int_{\max \tilde{y}}^{\max D} f(y)\, dy \simeq 0 \tag{5.8}$$

where $\tilde{y}$ denotes an empirically sampled observation.

Combining equations (5.5) and (5.7) yields the expression:

$$\ln O_k = \ln A_k + \varepsilon_{1,k} - \varepsilon_{2,k} \tag{5.9}$$

Substituting equation (5.6) into equation (5.9) gives:

$$\ln O_k = \ln (NW_k) + \ln f(y_k) + \varepsilon_{1,k} - \varepsilon_{2,k} \tag{5.10}$$

Finally, substituting equation (5.1) into equation (5.10) gives:

$$\ln O_k = \ln (NW_k) + \ln \left\{ \exp \left[\sum_{j=1}^{J} \theta_j \psi_j (y_k) - \eta \right] \right\} + \varepsilon_{1,k} - \varepsilon_{2,k} \tag{5.11}$$

which can be written more simply as:

$$\ln \left(\frac{O_k}{NW_k} \right) = -\eta + \sum_{j=1}^{J} \theta_j \psi_j (y_k) + \varepsilon_{1,k} - \varepsilon_{2,k} \tag{5.12}$$

Equation (5.12) can be rearranged into the matrix linear regression form:

$$\ln \left(\frac{O_k}{NW_k} \right) = \Psi\theta + \varepsilon \tag{5.13}$$

where Ψ is a nonstochastic matrix composed by horizontally concatenating a unit vector with the column vectors $\psi_j(y_k)\ \forall j$, θ is now amended to included the constant term and $\varepsilon = \varepsilon_{1,k} - \varepsilon_{2,k}$. Ordinary least squares parameter estimates of θ, in equation (5.13), are those of the LSFE. The need to apply numerical integration to calculate η (see equation (5.2)) is overcome as it is estimated as the negative of the constant term. This feature constitutes the primary computational advantage of the technique. Given that the econometric properties of the LSFE and its generalisations depend upon the characteristics of ε, the following section examines the properties of the error terms $\varepsilon_{1,k}$ and $\varepsilon_{2,k}$, respectively. For simplicity, the remainder of the chapter assumes that class widths are constant so that $W_k = W$, $\forall k$.

5.2 Error Properties

The previous section demonstrated that the LSFE error term is a composite of a quadrature error term and an empirical error term. This section examines the properties of these terms to derive unbiasedness and consistency conditions and the LSFE covariance matrix. Subsection 5.2.1 derives a formal expression of the LSFE quadrature error. Using this analysis, subsection 5.2.2 provides a numerical example of the size of the quadrature error for a bimodal generalised gamma density. Properties of the empirical or sampling error are given in subsection 5.2.3.

5.2.1 Derivation of Quadrature Error (ε_1)

Equation (5.5) in Section 5.1 defines the LSFE quadrature error for the kth class interval as:

$$\varepsilon_{1,k} = \ln E_k - \ln A_k \tag{5.14}$$

Given equations (5.3) and (5.6) this expression can be written more formally as:

$$\varepsilon_{1,k} = \ln\left\{\frac{\int_{a_k}^{a_k+W} f(y,\theta)\,dy}{Wf(y_k,\theta)}\right\} \tag{5.15}$$

which is independent of the sample size N. An important characteristic of $\varepsilon_{1,k}$ is that it is nonstochastic and as such $E(\varepsilon_{1,k}) = \varepsilon_{1,k}$.

Using a polynomial interpolation error formula of the $E_k \simeq A_k$ approximation outlined in Krommer and Ueberhuber (1994, p.74) it is possible to demonstrate that:

$$A_k - E_k = -\frac{NW^3}{24} f''(\xi_k,\theta), \qquad \xi_k \in (a_k, a_k+W) \tag{5.16}$$

Substituting equation (5.16) into equation (5.14) gives the LSFE quadrature error as

$$\begin{aligned} \varepsilon_{1,k} &= \ln\left(A_k + \frac{NW^3}{24} f''(\xi_k, \theta)\right) - \ln A_k \\ &= \ln\left(1 + \frac{NW^3 f''(\xi_k, \theta)}{24 A_k}\right) \end{aligned} \tag{5.17}$$

Substituting equation (5.6) into equation (5.17) gives the final expression:

$$\varepsilon_{1,k} = \ln\left(1 + \frac{W^2 f''(\xi_k, \theta)}{24 f(y_k, \theta)}\right) \tag{5.18}$$

Although equation (5.18) will typically be non-zero, the following limiting arguments can be invoked:

$$\lim_{W \to 0} \varepsilon_{1,k} \equiv \lim_{K \to \infty} \varepsilon_{1,k} = \mathbf{0} \tag{5.19}$$

Given that the generalised exponential family of distributions, for some random variable, y, have the form $f(y, \theta) = \exp[g(y, \theta) - \eta]$, where equation (5.1) demonstrates $g(y, \theta) = \sum_{j=1}^{J} \theta_j \psi_j(y)$, the first and second partial derivatives of $f(y, \theta)$ with respect to y are expressed as:

$$\begin{aligned} f'(y, \theta) &= g'(y, \theta) f(y, \theta) \\ f''(y, \theta) &= g''(y, \theta) f(y, \theta) + g'(y, \theta) f'(y, \theta) \\ &= \left\{ g''(y, \theta) + g'(y, \theta)^2 \right\} f(y, \theta) \end{aligned} \tag{5.20}$$

Applying equation (5.20) to equation (5.18) yields an alternate expression of $\varepsilon_{1,k}$:

$$\varepsilon_{1,k} = \ln\left(1 + \frac{W^2}{24} \left\{ g''(\xi_k, \theta) + g'(\xi_k, \theta)^2 \right\} \exp\left[g(\xi_k, \theta) - g(y_k, \theta)\right]\right) \tag{5.21}$$

which is independent of the normalising constant, η, found in $f(y, \theta)$. A special case of equation (5.21) is generated by assuming that $\xi_k = y_k$ (i.e., ξ_k is evaluated at class midpoints), to give:

$$\varepsilon_{1,k} \simeq \ln\left[1 + \frac{W^2}{24} \left\{ g''(y_k, \theta) + g'(y_k, \theta)^2 \right\}\right] \tag{5.22}$$

For given estimates, $\hat{\theta}$, equation (5.22) provides a convenient form for estimating the size of the quadrature errors.

5.2.2 Numerical Example

In order to examine the characteristics of $\varepsilon_{1,k}$ derived in subsection 5.2.1 a numerical example is constructed using the generalised gamma distribution. The density has a flexible form, capable of generating bimodality and contains as special cases the gamma, exponential, Weibull and power distributions. The unconditional generalised gamma density, $f(y)$, for some random variable y is written:

$$f(y) = \exp\left[\theta_1 \ln(y) + \theta_2 y + \theta_3 y^2 + \theta_4 y^3 - \eta\right], \quad 0 < y < \infty \tag{5.23}$$

where the normalizing constant, η, is defined as:

$$\eta = \ln \int_0^\infty \exp\left[\theta_1 \ln(y) + \theta_2 y + \theta_3 y^2 + \theta_4 y^3\right] dy \tag{5.24}$$

The distributional parameters $\theta = (22.5, -34.5, 6.75, -0.5)$ are selected in order to generate modes at $y = (1, 5)$ and an antimode at $y = 3$. The sample size and number of class intervals examined are $N = 2000$ and $K = 15$, respectively. Values for $\min \tilde{y}$ and $\max \tilde{y}$ for this density are selected such that

$$0.001 = \int_0^{\min \tilde{y}} f(y, \theta)\, dy = \int_{\max \tilde{y}}^{\infty} f(y, \theta)\, dy \tag{5.25}$$

to give $\max \tilde{y} = 6.3220$ and $\min \tilde{y} = 0.4640$; see equation (5.8). The value 0.001 is used to ensure Np_k does not approach zero and subsequently cause numerical instability. Components of the LSFE quadrature error, $\varepsilon_{1,k} = \ln(E_k/A_k)$, along with its approximation from equation (5.22), denoted $\hat{\varepsilon}_{1,k}$, are presented in Table 5.1.

Three features are apparent from Table 5.1. First, $\hat{\varepsilon}_{1,k}$ offers a reasonable approximation to $\varepsilon_{1,k}$. This demonstrates support for the assumption $\xi_k = y_k$. Secondly, the absolute magnitude of the error $\varepsilon_{1,k}$ is small relative to $\ln E_k$ and $\ln O_k$. As such, the majority of estimation error is likely to be generated by empirical sampling, $\varepsilon_{2,k}$. Finally, both $\varepsilon_{1,k}$ and $\hat{\varepsilon}_{1,k}$ increase as $k \to 1$ and $k \to K$, indicating that quadrature error is relatively larger over the tail domain of a distribution.[1]

[1]Estimators were constructed in Bakker (1995a, b) which employed knowledge of the quadrature error structure. However, given its relatively small magnitudes only minor gains in accuracy were achieved and are thus not reported in this chapter.

Table 5.1: Numerical example of quadrature error

$\ln E_k$	$\ln A_k$	$\varepsilon_{1,k}$	$\hat{\varepsilon}_{1,k}$
5.5866	5.4613	0.1254	0.1297
6.4610	6.5255	-0.0645	-0.0641
5.8384	5.8170	0.0214	0.0301
4.8326	4.7903	0.0423	0.0475
3.9684	3.9371	0.0313	0.0341
3.4381	3.4202	0.0179	0.0195
3.2827	3.2583	0.0244	0.0126
3.4109	3.3978	0.0131	0.0115
3.7570	3.7437	0.0134	0.0106
4.1844	4.1764	0.0080	0.0058
4.5571	4.5604	-0.0033	-0.0041
4.7340	4.7495	-0.0155	-0.0145
4.5724	4.5902	-0.0178	-0.0150
3.9302	3.9236	0.0066	0.0116
2.6663	2.5868	0.0795	0.0857

5.2.3 Empirical Error Term (ε_2)

The second component of the LSFE error term, namely $\varepsilon_{2,k}$, is defined by equation (5.7), and represents sampling error. Given that an observation can fall into one of K groups, where $\sum_k O_k = N$ and $\sum_k p_k = 1$, it is appropriate to model O_k as a multinomial random variable. Therefore, in vector notation:

$$\begin{aligned} E[O] &= N\mathbf{p} \\ var[O] &= N(\mathbf{D}_p - \mathbf{p}'\mathbf{p}) \end{aligned}$$

where $\mathbf{D}_p$ denotes the diagonal matrix based on the $(1 \times K)$ vector of probabilities $\mathbf{p} = (p_1, ..., p_K)$; see Bishop (1975, p.469). As a result:

$$E(\varepsilon_{2,k}) = E[\ln E_k - \ln O_k] = \mathbf{0} \tag{5.26}$$

In order to calculate the covariance matrix of $\varepsilon_{2,k}$, a first-order Taylor series expansion of $\varepsilon_{2,k} = \ln(E_k) - \ln(O_k)$ is applied as follows:

$$var(\varepsilon_{2,k}) = E\left(\varepsilon_{2,k}\varepsilon_{2,k}'\right) \simeq \mathbf{\Upsilon\Sigma\Upsilon}' \tag{5.27}$$

where:

$$\mathbf{\Upsilon}' = \begin{bmatrix} \partial\varepsilon_{2,1}/\partial O_1 & \partial\varepsilon_{2,1}/\partial O_2 & \cdots & \partial\varepsilon_{2,1}/\partial O_K \\ \partial\varepsilon_{2,2}/\partial O_1 & \partial\varepsilon_{2,2}/\partial O_2 & \cdots & \partial\varepsilon_{2,2}/\partial O_K \\ \vdots & & \ddots & \vdots \\ \partial\varepsilon_{2,K}/\partial O_1 & \partial\varepsilon_{2,K}/\partial O_2 & \cdots & \partial\varepsilon_{2,K}/\partial O_K \end{bmatrix}$$

$$= \begin{bmatrix} -O_1^{-1} & 0 & \dots & 0 \\ 0 & -O_2^{-1} & & \vdots \\ \vdots & & \ddots & 0 \\ 0 & \dots & 0 & -O_K^{-1} \end{bmatrix} \tag{5.28}$$
$$= \boldsymbol{\Upsilon}$$

and:

$$\begin{aligned} \boldsymbol{\Sigma} &= E\left[(O_k - E_k)(O_k - E_k)'\right] \\ &= N(\mathbf{D}_p - \mathbf{p}'\mathbf{p}) \end{aligned} \tag{5.29}$$

as O_k is multinomially distributed. Using sample data:

$$\begin{aligned} \hat{\boldsymbol{\Sigma}} &= N(\mathbf{D}_{\hat{p}} - \hat{\mathbf{p}}'\hat{\mathbf{p}}) \\ &= \frac{1}{N}\begin{bmatrix} O_1N - O_1^2 & O_1O_2 & \dots & O_1O_K \\ O_2O_1 & O_2N - O_2^2 & & \\ \vdots & & \ddots & \vdots \\ O_KO_1 & & \dots & O_KN - O_K^2 \end{bmatrix} \end{aligned} \tag{5.30}$$

where $\hat{\mathbf{p}} = O_k/N$. Therefore, the matrix approximation of $var(\varepsilon_{2,k})$ has the following toeplitz structure:

$$\begin{aligned} var(\varepsilon_{2,k}) &\simeq \begin{bmatrix} \frac{\hat{\Sigma}_{11}}{O_1^2} & \frac{\hat{\Sigma}_{12}}{O_1O_2} & \dots & \frac{\hat{\Sigma}_{1K}}{O_1O_K} \\ \frac{\hat{\Sigma}_{21}}{O_2O_1} & \frac{\hat{\Sigma}_{22}}{O_2^2} & & \\ \vdots & & \ddots & \vdots \\ \frac{\hat{\Sigma}_{K1}}{O_KO_1} & & \dots & \frac{\hat{\Sigma}_{KK}}{O_K^2} \end{bmatrix} \\ &= \frac{1}{N}\begin{bmatrix} \left(\frac{N}{O_1} - 1\right) & 1 & \dots & 1 \\ 1 & \left(\frac{N}{O_2} - 1\right) & & \vdots \\ \vdots & & \ddots & 1 \\ 1 & \dots & 1 & \left(\frac{N}{O_K} - 1\right) \end{bmatrix} \end{aligned} \tag{5.31}$$

Equation (5.31) demonstrates that the classical assumption of homoscedasticity is not supported by the LSFE. Therefore, the LSFE is inefficient. Improving the efficiency of the LSFE by using equation (5.31) in a generalised least squares framework is discussed in the following section.

5.3 Econometric Properties

Given the formal description of the LSFE error properties in Section 5.2 it is relatively straightforward to derive the conditions required for unbiasedness and consistency of the LSFE as discussed in subsection 5.3.1. Subsection 5.3.2 introduces a more efficient generalised least squares frequency estimator (GLSFE) that utilises the covariance matrix of the LSFE. Finally, subsection 5.3.3 includes Monte Carlo experiments designed to examine the efficiency gains of the GLSFE over the LSFE.

5.3.1 Unbiasedness and Consistency of the LSFE

Using standard ordinary least squares results it is possible to show that the sample parameter estimates, $\hat{\theta}$, of equation (5.13) are a linear function of θ in the following manner:

$$\hat{\theta} = \theta + (\Psi'\Psi)^{-1} \Psi' (\varepsilon_{1,k} - \varepsilon_{2,k}) \tag{5.32}$$

As $\varepsilon_{1,k}$ is nonstochastic $\varepsilon_{1,k} = E(\varepsilon_{1,k})$ (see subsection 5.2.1), and $E(\varepsilon_{2,k}) = \mathbf{0}$ (see subsection 5.2.3), the bias of the LSFE is derived as:

$$bias\left(\hat{\theta}\right) = E\left(\hat{\theta}\right) - \theta = (\Psi'\Psi)^{-1} \Psi'\varepsilon_{1,k} \tag{5.33}$$

Therefore, given that $\varepsilon_{1,k} \neq \mathbf{0}$, as demonstrated in Table 5.1, the LSFE is biased. However, employing the limiting argument of equation (5.19), gives:

$$\lim_{K\to\infty} bias\left(\hat{\theta}\right) = \mathbf{0} \tag{5.34}$$

Hence, parameter estimates approach unbiasedness as K increases.

Given the unbiasedness of $\hat{\theta}$ for large values of K, it is sufficient to prove consistency by showing convergence in mean square:

$$\lim_{N\to\infty} var\left(\hat{\theta}\right) = \mathbf{0} \tag{5.35}$$

The LSFE covariance matrix is derived from the basic definition:

$$var\left(\hat{\theta}\right) = E\left[\hat{\theta} - E\left(\hat{\theta}\right)\right]\left[\hat{\theta} - E\left(\hat{\theta}\right)\right]' \tag{5.36}$$

where equation (5.32) demonstrates:

$$\hat{\theta} - E\left(\hat{\theta}\right) = (\Psi'\Psi)^{-1} \Psi' \left\{\varepsilon_{1,k} - \varepsilon_{2,k} - E(\varepsilon_{1,k}) + E(\varepsilon_{2,k})\right\} \tag{5.37}$$

As $\varepsilon_{1,k}$ is nonstochastic $\varepsilon_{1,k} = E(\varepsilon_{1,k})$ (see Subsection 5.2.1), and $E(\varepsilon_{2,k}) = \mathbf{0}$ (see subsection 5.2.3), equation (5.37) can be written more simply as:

$$\hat{\theta} - E\left(\hat{\theta}\right) = -(\Psi'\Psi)^{-1} \Psi'\varepsilon_{2,k} \tag{5.38}$$

Applying equation (5.38) to equation (5.36) gives the LSFE covariance matrix as:

$$var\left(\hat{\theta}\right) = \left(\Psi'\Psi\right)^{-1} \Psi'\Omega\Psi \left(\Psi'\Psi\right)^{-1} \tag{5.39}$$

where $\Omega = E\left(\varepsilon_{2,k}\varepsilon_{2,k}'\right)$. As Ψ is independent of N, the proof of $\hat{\theta}$ consistency reduces to demonstrating:

$$\lim_{N\to\infty} \Omega = \mathbf{0} \tag{5.40}$$

Using the equation (5.31) approximation of $E\left(\varepsilon_{2,k}\varepsilon_{2,k}'\right)$ yields:

$$\Omega \simeq \frac{1}{N}\begin{bmatrix} \left(\frac{N}{O_1}-1\right) & 1 & \cdots & 1 \\ 1 & \left(\frac{N}{O_2}-1\right) & & \vdots \\ \vdots & & \ddots & 1 \\ 1 & \cdots & 1 & \left(\frac{N}{O_K}-1\right) \end{bmatrix} \tag{5.41}$$

The off-diagonal elements, $(1/N)$, tend to zero as N increases and the diagonal elements:

$$Diag\left(\Omega\right) = \frac{1}{O_k} - \frac{1}{N}$$

also tend to zero as N increases given that $\sum_k O_k = N$. Hence equation (5.40) is satisfied and the LSFE exhibits consistency for sufficiently large K.

Observe that as $N \to \infty$, holding K constant, $\Omega \to \mathbf{0}$ so that $var\left(\hat{\theta}\right) \to \mathbf{0}$. However, $E\left(\hat{\theta}\right) \to \theta$ only as $K \to \infty$, while holding N constant, results in $O_k \to \mathbf{0}$, $Diag\left(\Omega\right) \to \infty$ and $var\left(\hat{\theta}\right) \to \infty$. Thus, as K increases the expected value of parameter estimates approach their true population values. However, their associated variance also increase. The trade-off between unbiasedness and efficiency is further examined in Section 5.4.

5.3.2 Generalised Least Squares Frequency Estimator

Given that Ω, as defined by equation (5.41), is not equal to $\sigma^2 I$ (where I is the identity matrix), the LSFE covariance matrix is not of the classical form $\sigma^2\left(\Psi'\Psi\right)^{-1}$ and the LSFE is inefficient. Efficient estimation of θ requires application of the covariance matrix Ω in the generalised least squares (GLS) framework; see Greene (1990, p.385). Given that Ω is a known, symmetric, positive definite matrix it is possible to derive the GLS estimates as:

$$\theta^* = \left(\Psi'\Omega^{-1}\Psi\right)^{-1} \Psi'\Omega^{-1} \left(\ln O\right) \tag{5.42}$$

where:

$$var\left(\theta^*\right) = \left(\Psi'\Omega^{-1}\Psi\right)^{-1} \tag{5.43}$$

The parameter estimates, θ^*, are defined as belonging to the generalised least squares frequency estimator (GLSFE).

Given that the GLS analogue of equation (5.32) possesses the form:

$$\hat{\theta}^* = \theta + \left(\Psi'\Omega^{-1}\Psi\right)^{-1}\Psi'\Omega^{-1}\left(\varepsilon_1 - \varepsilon_2\right)$$

it is possible to derive the biasedness of the GLSFE as:

$$bias\left(\hat{\theta}^*\right) = \left(\Psi'\Omega^{-1}\Psi\right)^{-1}\Psi'\Omega^{-1}\varepsilon_1 \tag{5.44}$$

where $\varepsilon_{1,k} = E\left(\varepsilon_{1,k}\right)$ given $\varepsilon_{1,k}$ is nonstochastic (see Subsection 5.2.1), and $E\left(\varepsilon_{2,k}\right) = \mathbf{0}$ (see subsection 5.2.3). Employing the limiting argument of equation (5.19), gives:

$$\lim_{K\to\infty} bias\left(\hat{\theta}^*\right) = \mathbf{0} \tag{5.45}$$

Hence, parameter estimates approach unbiasedness as K increases. Assuming $\varepsilon_{1,k}$ is negligible from sufficiently large K, the GLSFE estimator, θ^*, is the minimum variance linear unbiased estimator in the generalised regression model by application of the Gauss-Markov theorem.

5.3.3 Monte Carlo Experiments

This subsection employs Monte Carlo techniques to examine the efficiency gains of the GLSFE over the LSFE. The estimators are applied to random samples generated from the generalised gamma distribution as discussed in subsection 5.2.2. Generalised gamma random numbers are generated using the inverse cumulative density method. The following steps are adopted. For values given in θ calculate the normalising constant, η, such that $\int_0^\infty f\left(y\right)dy = 1$. Using the probability density function, $f(y)$, derive the corresponding cumulative density function (cdf). As the cdf is defined over $[0, 1]$ a uniformly distributed random variable, defined over this interval, is used to select the value of y that would have generated such a cdf value. The random number generator seed value of 12345678 is used at the beginning of each Monte Carlo experiment.

In chapter 2, Creedy, Lye and Martin found that distributions with long tails, such as the generalised gamma distribution, are likely to generate extreme observations. These values generate groupings of the frequency classes over less important tail regions of the density surface and subsequently result in higher aggregation of classes around modal regions. Hence, sample estimates are likely to be biased by the presence of outliers. Unreported experiment results by the present author corroborate this finding. Therefore, in all experiments the influence of outliers is reduced by simulating $1.05N$ observations, sorting the data and removing the $0.05N$ largest observations.

The distributional parameters $\theta = \{40.5, -40.5, 6.75, -0.5\}$ applied to equation (5.23), which generate a symmetrical bell-shaped density with a single mode at $y = 3$, are used to generate random samples.[2] Monte Carlo experiments employ 1000 replications, sample sizes of $N = (500, 2000)$ and class intervals of $K = (10, 30)$. Given the use of a common random number generator seed, the same random samples are applied. For comparative purposes, statistics are calculated for both the LSFE and GLSFE. The mean of parameter estimates, their standard deviation (SD) and root mean square error (RMSE) are presented in Table 5.2.

The results demonstrate that $K = 10$ is too small to characterise adequately the simulated distribution as parameter estimates are clearly biased and inconsistent. However, the standard deviations of parameter estimates from the GLSFE are clearly superior to those of the LSFE. This supports the claim that the GLSFE is efficient relative to the LSFE. Evidence as to which estimator is more 'accurate' is mixed. In general, there appears to be a positive relationship between N and the optimal number of class intervals, K^*. For example, LSFE results for $(K = 10, N = 500)$ appear more accurate than $(K = 30, N = 500)$. Results in Bakker (1995a) suggest bimodal densities shift the relationship such that a higher K^* is associated with a given N. The nature of this relationship is more formally examined in the following section.

5.4 Optimal (K,N) Relationship

For a sample of size N, it is of interest to determine the optimal number of class intervals, K^*, to apply during estimation of the LSFE and GLSFE. The issue involves a trade-off between quadrature error, $\varepsilon_{1,k}$, and sampling error, $\varepsilon_{2,k}$. As K increases, the bias from quadrature error decreases; however, the variation from sampling an empirical distribution with smaller class widths subsequently increases. Subsection 5.4.1 advocates the trace of the mean squared error matrix as a criterion to be minimised when determining the optimal trade-off between the biasedness and variance of LSFE and GLSFE. A numerical approximation to the criterion is given in subsection 5.4.2. Finally, subsection 5.4.3 contains a numerical example that provides estimates of optimal (K, N) schedules for the LSFE and GLSFE across three

[2] Other densities considered include mode plus inflection point and bimodal specifications. However, results are qualitatively similar and available in Bakker (1995a).

Table 5.2: Comparison of LSFE and GLSFE

K	N		$\theta_1 = 40.5$	$\theta_2 = -40.5$	$\theta_3 = 6.75$	$\theta_4 = -0.5$
10	500	Mean LSFE	46.6350	-43.1568	6.6934	-0.4643
		S.D. LSFE	23.3418	25.5438	4.4535	0.3314
		RMSE LSFE	24.1233	25.6689	4.4517	0.3332
		Mean GLSFE	48.8950	-45.8258	7.1839	-0.5021
		S.D. GLSFE	19.7183	21.4957	3.7501	0.2801
		RMSE GLSFE	21.4219	22.1352	3.7733	0.2800
	2000	Mean LSFE	46.5860	-43.0793	6.6764	-0.4627
		S.D. LSFE	13.0429	14.3215	2.4983	0.1856
		RMSE LSFE	14.3870	14.5448	2.4981	0.1892
		Mean GLSFE	48.4169	-45.1732	7.0502	-0.4908
		S.D. GLSFE	9.5445	10.4322	1.8251	0.1366
		RMSE GLSFE	12.3969	11.4264	1.8487	0.1369
30	500	Mean LSFE	32.0565	-29.8033	4.7200	-0.3407
		S.D. LSFE	18.5548	21.1871	3.8243	0.2927
		RMSE LSFE	20.3772	23.7248	4.3280	0.3331
		Mean GLSFE	38.0368	-36.9849	6.0362	-0.4404
		S.D. GLSFE	15.0250	16.9522	3.0432	0.2326
		RMSE GLSFE	15.2182	17.3045	3.1243	0.2400
	2000	Mean LSFE	39.5647	-37.7926	6.0586	-0.4355
		S.D. LSFE	10.9693	12.5664	2.2637	0.1723
		RMSE LSFE	11.0037	12.8486	2.3659	0.1839
		Mean GLSFE	42.2374	-41.2064	6.7134	-0.4870
		S.D. GLSFE	7.3141	8.2523	1.4812	0.1132
		RMSE GLSFE	7.5141	8.2784	1.4809	0.1139

benchmark scenarios.

5.4.1 The MSE Criterion

The mean squared error (MSE) matrix provides a convenient measure for analysing the trade-off between the biasedness and variance of an estimator. The MSE matrix is defined as:

$$\begin{aligned} MSE\left[\hat{\theta}\right] &= E\left[\left(\hat{\theta}-\theta\right)\left(\hat{\theta}-\theta\right)'\right] \\ &= E\left[\hat{\theta}-E\left(\hat{\theta}\right)\right]\left[\hat{\theta}-E\left(\hat{\theta}\right)\right]'+\left[E\left(\hat{\theta}\right)-\theta\right]\left[E\left(\hat{\theta}\right)-\theta\right]' \\ &= var\left(\hat{\theta}\right)+\left[bias\left(\hat{\theta}\right)\right]\left[bias\left(\hat{\theta}\right)\right]' \end{aligned} \tag{5.46}$$

The trace of $MSE\left[\hat{\theta}\right]$ is commonly used as a scalar representation:

$$tr\left\{MSE\left[\hat{\theta}\right]\right\}=\sum_j\left\{var\left(\hat{\theta}_j\right)+\left[bias\left(\hat{\theta}_j\right)\right]^2\right\} \tag{5.47}$$

This measure effectively attaches equal weight to the variance and biasedness of an estimator. By using the trace of $MSE\left[\hat{\theta}\right]$ as a cost function, the optimal number of class intervals, K^*, for a given sample size can be expressed as the solution of

$$K^*=\min_K tr\left\{MSE\left[\hat{\theta}\right]\middle| N\right\} \tag{5.48}$$

Equation (5.47) demonstrates that application of the MSE criterion requires explicit functional form for $bias\left(\hat{\theta}\right)$ and $var\left(\hat{\theta}\right)$. The biasedness and covariance matrices of the LSFE are given in equations (5.33) and (5.39) respectively as:

$$\begin{aligned} bias\left(\hat{\theta}\right) &= \left(\Psi'\Psi\right)^{-1}\Psi'\varepsilon_1 \\ var\left(\hat{\theta}\right) &= \left(\Psi'\Psi\right)^{-1}\Psi'\Omega\Psi\left(\Psi'\Psi\right)^{-1} \end{aligned}$$

where ε_1 can be calculated numerically for a given θ using equation (5.15) and Ω is defined by equation (5.41). The biasedness and covariance matrices of the GLSFE are given in equations (5.44) and (5.43), respectively as:

$$\begin{aligned} bias\left(\hat{\theta}^*\right) &= \left(\Psi'\Omega^{-1}\Psi\right)^{-1}\Psi'\Omega^{-1}\varepsilon_1 \\ var\left(\hat{\theta}^*\right) &= \left(\Psi'\Omega^{-1}\Psi\right)^{-1} \end{aligned}$$

However, given K^* is not continuous with respect to K, equation (5.48) is not differentiable. Therefore, an alternative numerical approach is explained below.

5.4.2 A Numerical Solution

Given that equation (5.48) defines K^* as a noncontinuous function with respect to K, it is not possible to apply a calculus solution. Therefore, this section employs continuous approximations to $var\left(\hat{\theta}_j\right)$ and $bias\left(\hat{\theta}_j\right)$, thus enabling a solution from first order conditions. Numerical analysis in subsection 5.4.3 indicates that $\sum_j \left[bias\left(\hat{\theta}_j\right)\right]^2$ exhibits exponential decay as K increases, *ceteris paribus*. Therefore, a convenient approximation is:

$$\sum_j \left[bias\left(\hat{\theta}_j\right)\right]^2 \simeq \exp\left[\delta_1 + \delta_2 K\right] \tag{5.49}$$

where δ_1 and δ_2 are considered functions of $\theta, \min \tilde{y}$ and $\max \tilde{y}$. Observe that equation (5.49) is independent of N and that $bias\left(\theta_j\right)$ can be calculated without having to employ simulation methods as $\varepsilon_{1,k}$ in equations (5.33) and (5.44) can be calculated analytically for a given θ. In addition, results from subsection 5.4.3 demonstrate that the trace of the covariance matrix, $\sum_j var\left(\hat{\theta}_j\right)$, is well approximated by the functional form:

$$\sum_j var\left(\hat{\theta}_j\right) \simeq \beta_1 + \beta_2\left(\frac{1}{N}\right) + \beta_3\left(\frac{K}{N}\right) \tag{5.50}$$

where β_1 and β_2 are considered functions of $\theta, \min \tilde{y}$ and $\max \tilde{y}$. It is also possible to avoid simulation methods when calculating $var\left(\theta_j\right)$ by replacing observed frequencies in Ω, defined in equation (5.41), with expected values. By substituting equations (5.49) and (5.50) into equation (5.47) enables the minimum solution to equation (5.48) to be derived from the first order condition as:

$$K^* \simeq \gamma_1 + \gamma_2 \ln N \tag{5.51}$$

where

$$\gamma_1 = \frac{\ln\left(-\beta_2/\delta_2\right) - \delta_1}{\delta_2}$$

$$\gamma_2 = -\frac{1}{\delta_2}$$

The second order condition demonstrates that K^* minimises equation (5.48) regardless of the signs on δ and β.

Table 5.3: Generalised gamma PDF modes and associated parameters

Scenario	Modes	θ_1	θ_2	θ_3	θ_4
I	$(3, 3, 3)$	40.5	-40.5	6.75	-0.5
II	$(1, 3, 3)$	13.5	-22.5	5.25	-0.5
III	$(1, 3, 5)$	22.5	-34.5	6.75	-0.5

The usefulness of equation (5.51) may be questioned, given that an empirical researcher has no *a priori* knowledge of θ. However, valuable insight into the theoretical properties of the estimator are gained by considering distributional benchmark scenarios: unimodality (I), mode plus inflection point (II) and bimodality (III). Subsection 5.4.3 applies this analysis to both the LSFE and GLSFE.

5.4.3 Numerical Example

This section applies the optimal (K, N) analysis from the previous subsection to the generalised gamma density, $f(y)$, for both the LSFE and GLSFE. Turning points in $f(y)$ have been selected in order to generate three distributional types. Scenario I has a unimodal specification with mode at $y = 3$; Scenario II has mode and inflection point at $y = 1$ and $y = 3$ respectively; Scenario III has a bimodal specification with modes at $y = 1, 5$ and an antimode at $y = 3$. These specifications are selected as benchmarks for the simplest to most complicated type of form that the generalised gamma density can generate. The associated parameters are shown in Table 5.3. Values for $\min \tilde{y}$ and $\max \tilde{y}$ of these densities are selected in order to satisfy equation (5.25) in subsection 5.2.2.

Ordinary least squares estimates of δ_1 and δ_2, from equation (5.49), over the domain $20 \leq K \leq 60$, are calculated as:

$$\sum_j \left[bias\left(\hat{\theta}_j\right)\right]^2 = \exp\left[\underset{(0.0887)}{2.4407} - \underset{(0.0021)}{0.0696}\,K\right]$$

where standard errors are presented in parenthesis and $\bar{R}^2 = 0.964$. Similarly high degrees of fit for equation (5.49) are observed across all scenarios and for both the LSFE and GLSFE. For $20 \leq K \leq 60$ and $N = (500, 600, \ldots, 1500)$ ordinary least squares parameter estimates of β, from equation (5.50), are calculated as:

$$\sum_j var\left(\hat{\theta}_j\right) = \underset{(0.4057)}{18.6567} - \underset{(0.0002)}{0.0082}\left(\frac{1}{N}\right) + \underset{(4.3705)}{104.1274}\left(\frac{K}{N}\right)$$

Table 5.4: Optimal (K,N) relationship for LSFE

Scenario	LSFE		GLSFE	
	$\hat{\gamma}_1$	$\hat{\gamma}_2$	$\hat{\gamma}_1$	$\hat{\gamma}_2$
I	-40.3618	7.1618	-33.2763	7.4114
II	-59.0390	11.1854	-29.8930	8.7397
III	-70.0123	14.3750	-30.6726	10.9474

where $\bar{R}^2 = 0.927$. Again, similarly high degrees of fit for equation (5.50) are observed across all scenarios and for both the LSFE and GLSFE. From the estimates of $\hat{\delta}$ and $\hat{\beta}$ it is possible to solve for $\hat{\gamma}$, using equation (5.51). LSFE and GLSFE estimates of $\hat{\gamma}$ for scenarios I, II and III are presented in Table 5.4. While $N = (500, 600, \ldots, 1500)$ is held constant across scenarios, for numerical stability estimates of scenarios I and II are derived over the domain $10 \leq K \leq 50$.

The results indicate that as the shape of a density becomes more complicated (e.g., approaches bimodality), $\hat{\gamma}_2$ increases and for the LSFE $\hat{\gamma}_1$ becomes smaller. Therefore, the intuitive result that a simple distribution will require fewer class intervals than a more complicated to characterise its shape one is verified. For example, with a sample of $N = 2000$ observations the optimal number of class intervals to apply to the LSFE for scenarios I, II and III respectively is estimated as $K^* = (14, 26, 39)$ and for the GLSFE as $K^* = (23, 37, 53)$. Finally, holding other things equal, it is optimal for the GLSFE to apply a larger number of class intervals, for a given sample size, than the LSFE.

5.5 LSFE and GLSFE Inference

Creedy *et al.* (1996a) warn that the sampling theory of the LSFE has not been established, so that estimated standard errors should be treated with caution. In general terms, this section attempts to address the issue of whether the estimated t-statistics from the GLSFE can be validly interpreted. Subsection 5.5.1 outlines the assumptions required to conduct conventional hypothesis testing. The chi-squared goodness-of-fit statistic is introduced as a means of testing these assumptions. Subsection 5.5.2 uses Monte Carlo simulation to determine whether the GLSFE satisfies the chi-squared goodness-of-fit statistic.

5.5.1 Inference Assumptions

Statistical inference for the linear regression model, $\ln O = \Psi\theta + \varepsilon$, is based on three crucial assumptions: (i) nonstochastic Ψ, (ii) knowledge of the covariance matrix and (iii) normally distributed errors for small samples. However, it is possible to show under general conditions that $\hat{\theta}$ is asymptotically normal regardless of the error term's distribution. Under these assumptions:

$$t_j = \frac{\hat{\theta}_j - \theta_j}{\sqrt{diag\left\{var\left(\hat{\theta}\right)\right\}_j}} \tag{5.52}$$

is distributed as the Student's t distribution with $K-J-1$ degrees of freedom.

As the 't-statistic' in equation (5.52) is the basis of conventional hypothesis testing, the estimated t-statistics should posses its distributional properties. A method of testing this proposition is to construct an empirical distribution of estimated t-statistics, using Monte Carlo simulation, and employ the chi-squared goodness-of-fit test. The test statistic examines whether the empirical distribution differs significantly from a Student's t distribution with $K-J-1$ degrees of freedom and is calculated using:

$$\chi^2 = \sum_k \frac{\left(f_{o,k} - f_{t,k}\right)^2}{f_{t,k}} \tag{5.53}$$

where k is the number of class intervals, $f_{o,k}$ is the empirical distribution's observed frequency in the kth class and $f_{t,k}$ the theoretical frequency (calculated from the density specified under the null hypothesis) in the kth class. As the restriction $\sum_k f_{t,k} = \sum_k f_{o,k}$ is employed, the critical value is determined from a χ^2 density with $k-1$ degrees of freedom.

5.5.2 Monte Carlo Testing of Inference Assumptions

To determine whether the LSFE or GLSFE satisfies inference assumptions outlined in the previous section, an empirical distribution of t-statistics is generated using Monte Carlo simulation and tested using the chi-squared 'goodness-of-fit' test statistic, specified in equation (5.53), against the null of Student's $t_{df=K-J-1}$. For comparative purposes, the scenario II population density, outlined in Table 5.3, is used to generate empirical GLSFE t-statistic distributions, using the 'appropriate' equation (5.43) covariance matrix, and LSFE t-statistic distributions assuming the 'inappropriate' classical regression model covariance matrix $\sigma\left(\Psi'\Psi\right)^{-1}$. Random samples of $N = 2000$ observations are generated as described in subsection 5.3.3, and estimated

Table 5.5: χ^2 distribution test statistics

K	$\theta_1 = 13.5$	$\theta_2 = -22.5$	$\theta_3 = 5.25$	$\theta_4 = -0.5$
		LSFE		
25	11028635.83	115957.63	9250.26	2146.83
50	231185.86	6805.77	1097.47	549.63
75	7289.20	1153.85	714.18	593.70
100	2641.26	2166.52	2088.07	1156.97
125	21618.78	9537.92	7143.82	5529.98
		GLSFE		
25	14371.32	2253.84	575.18	152.73
50	1006.50	273.13	88.96	29.30**
75	157.80	45.07	17.55**	17.16**
100	27.60**	32.93*	58.41	69.09
125	73.65	109.30	160.86	151.67

* Indicates significance at the 1 per cent level.

** Indicates significance at the 5 per cent level.

using $K = (25, 50, 75, 100, 125)$; 10,000 replications are employed. Using $k = 20$ in (5.53), the χ^2 statistics are reported in Table 5.5.

The GLSFE test statistics are clearly superior to those of the LSFE. Moreover, the evidence clearly indicates that conventional hypothesis testing applied to the LSFE, assuming the covariance matrix $\sigma\left(\Psi'\Psi\right)^{-1}$, is invalid regardless of the value of K. However, t-statistics generated using the GLSFE covariance matrix appear satisfactory if estimation occurs using approximately twice the number of class intervals as suggested by the (K^*, N) relationship derived in Subsection 5.4.3 from the mean square error criterion.

5.6 Conclusions

This chapter formally derived the least squares frequency estimator (LSFE), introduced by Martin (1990), by expressing the natural logarithm of the 'true' frequency, for a particular density and class interval, using an empirical and numerical integration approximation. The estimator was shown to approach unbiasedness as the number of class intervals increases. Assuming the number of class intervals is 'sufficiently large' it is possible to demonstrate consistency. However, a fundamental trade-off exists between the biasedness and precision of the estimator. The covariance matrix of the LSFE was shown to differ from the standard classical regression model covariance matrix. As such, conventional hypothesis testing is not valid. Using the LSFE covariance matrix with the generalized least squares methodology, the corresponding GLSFE was shown to produce efficiency gains.

The optimal relationship between the number of class intervals and sample size for the LSFE and GLSFE was derived, using the mean square error criterion, for the benchmark scenarios of unimodality, mode plus inflection point and bimodal densities. The intuitive result that for a given sample size a simple distribution will require fewer class intervals to characterise its shape than a more complicated one was verified. As the sample size increases for a given distribution it is approximately optimal to increase the number of class intervals used for both estimators in a logarithmic manner. With a given sample size and distribution, it is optimal for the GLSFE to apply a larger number of class intervals than the LSFE. Monte Carlo results indicate that hypothesis testing with the GLSFE is valid if estimates are obtained using approximately twice the optimal number of class intervals derived by the mean square error criterion.

Chapter 6

Age and the Distribution of Earnings

Alex Bakker and John Creedy

This chapter uses the approach to modelling the income distribution that was developed in chapter 2, and applies the estimation methods for estimating the generalised exponential family of distributions with grouped data developed in chapter 5, in order to examine the changing distribution of earnings with age. The income distribution model involves both deterministic and stochastic components. Individuals' incomes are seen as resulting from a market supply and demand model where the reduced form represents the deterministic component that can generate multiple solutions. A stochastic error-correction mechanism is imposed and used to derive the income distribution. The model generates a flexible functional form described as the generalised gamma distribution. Special cases of this general distribution include the gamma, exponential, Weibull and power distribution.

The model is used to estimate the changing distribution of income with age for New Zealand males and females, using a special set of grouped data. The present approach can be contrasted with one in which a simple functional form is assumed to apply in each age group, where the arithmetic mean (of logarithms) is expressed as a function of age using a human capital argument, and a stochastic process describing relative income mobility is used in order to generate the changing variance (of logarithms) of income with age; see Creedy (1985). The approach used in this chapter allows for the possibility that the functional form itself can change over the life cycle as a result of demand and supply changes. This considerable increase in flexibility is obtained without a large increase in the number of parameters to be estimated.

The model, taken from chapter 2, is described briefly in Section 6.1. Estimation methods for grouped data are presented in Section 6.2. The empirical results are examined in Section 6.3.

6.1 A Demand and Supply Model

Demand and supply analysis is employed in subsection 6.1.1 to derive the deterministic model. Subsection 6.1.2 extends the deterministic model with stochastic components and derives the associated analytical distribution, following chapter 2.

6.1.1 The Deterministic Model

Consider a labour market in which all individuals are employed. The ith individual's income, y_i, is equal to the product of the wage rate per hour, w_i, and the number of hours worked, n_i. Thus:

$$y_i = w_i n_i \tag{6.1}$$

The demand for goods can be expressed as the following function of the wage:

$$x_i^d = \alpha_i - \frac{\beta_i}{w_i} \tag{6.2}$$

The term $1/w_i$ represents the price of goods in terms of labour, so that the associated labour supply function is given by:

$$n_i^s = \frac{x_i^d}{w_i} \tag{6.3}$$

Suppose the labour demand function involves the demand for labour being expressed as the following function of the wage:

$$n_i^d = \delta_i - \gamma_i w_i \tag{6.4}$$

Using equation (6.2) to express w_i in terms of x_i^d, so that $w_i = \beta_i/(\alpha_i - x_i^d)$, and substituting into equations (6.3) and (6.4) gives the quadratic labour supply and reciprocal labour demand curves, respectively, as:

$$n_i^s = \left(\frac{\alpha_i}{\beta_i}\right) x_i^d - \left(\frac{1}{\beta_i}\right) \left(x_i^d\right)^2 \tag{6.5}$$

$$n_i^d = \delta_i + \frac{\beta_i \gamma_i}{x_i^d - \alpha_i} \tag{6.6}$$

In equilibrium, the supply of labour given by equation (6.5) must equal the demand in equation (6.6). Writing $x_i = x_i^d$, and substituting for $x_i = y_i$, this equilibrium condition gives the following cubic for the ith individual in the market:

$$y_i^3 - 2\alpha_i y_i^2 + (\alpha_i^2 + \beta_i\delta_i)y_i - (\alpha_i\beta_i\delta_i - \gamma_i\beta_i^2) = 0 \tag{6.7}$$

6.1.2 The Stochastic Specification

Suppose the parameters of each individual's demand and supply functions are subject to stochastic shocks which lead to movements of income away from the equilibrium value. In the present model, equilibrium values are defined as the real roots of:

$$\mu_i(y_i) = y_i^3 - 2\alpha_i y_i^2 + (\alpha_i^2 + \beta_i\delta_i)y_i - (\alpha_i\beta_i\delta_i - \gamma_i\beta_i^2) \tag{6.8}$$

which is obtained from equation (6.7). Incomes are subsequently assumed to return or 'error-correct' to their values which held before the shocks. An autonomous Itô process represents a continuous time representation of such a stochastic process whereby the change in the ith individual's income is:

$$dy_i = -\mu_i(y_i)\,dt + \sigma_i(y_i)\,dW_i \tag{6.9}$$

Here $\mu_i(y_i)$ is the instantaneous mean, $\sigma_i^2(y_i)$ is the instantaneous variance and dW_i is the increment of a Wiener process, defined as $dW_i \equiv \lim_{\Delta t\to 0}\sqrt{\Delta t}Z_i = \sqrt{dt}Z_i$, where $Z_i \sim N(0,1)$, so that $dW_i(t) \sim N(0, dt)$. For further details of this type of process see chapter 1. Assuming that the instantaneous variance of the ith individual is proportional to income gives:

$$\sigma_i^2(y_i) = \tau_i y_i, \qquad \tau_i > 0 \tag{6.10}$$

Suppose that the random shocks applying to individuals are drawn from the same distribution and that individuals have the same demand and supply functions. These simplifications make it possible to drop the individual subscripts from equations (6.8) to (6.10) and regard the resulting distribution as applying over all individuals in the market. Heterogeneity can be introduced, as shown below, by allowing some of the structural parameters to be functions of various characteristics such as age.

It is shown in chapter 2 that this model leads to a stationary distribution of the form:

$$f_s(y) = \exp\left[\theta_1 \ln y + \theta_2 y + \theta_3 y^2 + \theta_4 y^3 - \eta\right], \qquad 0 < y < \infty \tag{6.11}$$

where:

$$\theta_1 = \frac{2(\alpha\beta\delta - \gamma\beta^2)}{\tau} - 1 \tag{6.12}$$

$$\theta_2 = \frac{-2(\alpha^2 + \beta\delta)}{\tau} \tag{6.13}$$

$$\theta_3 = \frac{2\alpha}{\tau} \tag{6.14}$$

$$\theta_4 = \frac{-2}{3\tau} \tag{6.15}$$

and:

$$\eta = \ln \int_0^\infty \exp\left[\theta_1 \ln y + \theta_2 y + \theta_3 y^2 + \theta_4 y^3\right] dy \tag{6.16}$$

The four distributional parameters $(\theta_1, \theta_2, \theta_3, \theta_4)$ are functions of the four structural parameters $(\alpha, \beta, \gamma, \delta)$ plus the variance coefficient τ and therefore cannot all be recovered. However, using equations (6.12) to (6.15) it is possible to calculate $(\tau, \alpha, \delta\beta, \gamma\beta^2)$. The emphasis here is on the form of the distribution and hence the distributional parameters θ, so the lack of identifiability of the structural parameters is not a serious problem.

The distribution given by equation (6.11) is the generalised gamma. If the economic model given by equations (6.1) to (6.4) contains a unique equilibrium the distribution is unimodal. Unlike the standard gamma distribution, proposed by Salem and Mount (1974) to describe income distributions, the generalised gamma can also be multimodal. Different specifications of the demand and supply functions, equations (6.4) and (6.5), respectively, give rise to alternative expressions for the exponential term in equation (6.11) by, for example, increasing or reducing the order of the polynomial in y. However, each is a special case of the exponential family.

The form given in equation (6.11) describes an unconditional distribution. However, the distributional parameters, θ, can in turn be expressed as functions of a set of exogenous variables which give rise to changes in the structural parameters. In the present context, the parameters are modelled as functions of age. This gives rise to a set of conditional distributions. The estimation of each case is examined in the following section.

6.2 Estimation Using Grouped Data

Income data are often unavailable in unit record form, and are only given in grouped (frequency) format, as in the present case. This section presents the estimation method for exponential family densities with data in frequency form, from chapter 5. The exponential family of unconditional density functions, $f(y)$, is defined for the random variable y as:

$$f(y) = \exp\left[\sum_{j=1}^{J} \theta_j \psi_j(y) - \eta\right], \qquad y \in D \tag{6.17}$$

where $\psi_j(.)$ is a function depending on the density type, θ_j are the associated distributional parameters, J is the number of additive terms in the density and η is the normalising constant given by:

$$\eta = \ln \int_D \exp\left[\sum_{j=1}^{J} \theta_j \psi_j(y)\right] dy \tag{6.18}$$

thus ensuring $\int_D f(y)\, dy = 1$. While θ_j terms are constant for unconditional analysis, conditional modelling expresses the jth distributional term for the ith individual, $\theta_{j,i}$, as a function of exogenous variables relevant to jth term, $z_{j,i}$, such that:

$$\theta_{j,i} = g_j(z_{j,i}) \tag{6.19}$$

where the function $g_j(.)$ is possibly nonlinear.

Subsection 6.2.1 outlines the generalised least squares frequency estimator (GLSFE), which is used to obtain estimates of unconditional distribution parameters. Subsection 6.2.2 demonstrates how the log-likelihood function associated with conditional exponential family densities can be amended for grouped data.

6.2.1 Unconditional Distribution Estimates

In chapter 5 it was shown that to use an equation of the form given in (6.17) in a frequency context it is necessary to divide the density's domain into $k = 1, ..., K$ class intervals, such that the probability associated with the kth class interval is:

$$p_k = \int_{a_k}^{a_k + W_k} \exp\left[\sum_{j=1}^{J} \theta_j \psi_j(y) - \eta\right] dy \tag{6.20}$$

where a_k denotes the lower bound of the kth class interval of width W_k. For a sample of size N, the natural logarithm of the expected frequency in the kth class interval is $\ln E_k = \ln(Np_k)$. The LSFE is derived from two approximations of $\ln E_k$.

The first approximation of $\ln E_k$ arises from computing the integral in equation (6.20) numerically. This expression can be written as:

$$\ln E_k = \ln A_k + \varepsilon_{1,k} \tag{6.21}$$

where $\varepsilon_{1,k}$ is the numerical integration or 'quadrature' error. The term A_k represents a rectangular area approximation of the integral in equation (6.20) and is calculated as:

$$A_k = NW_k f\left(y_k\right) \tag{6.22}$$

where $y_k = a_k + 0.5W_k$ is the midpoint of the kth class.

The second approximation of $\ln E_k$ represents an empirical approximation and is written:

$$\ln E_k = \ln O_k + \varepsilon_{2,k} \tag{6.23}$$

where O_k is the empirical frequency observed for the kth class interval and $\varepsilon_{2,k}$ is the empirical sampling error of the kth class interval. It is assumed that the sample size is sufficiently large such that:

$$\int_{\min D}^{\min \tilde{y}} f\left(y\right) dy + \int_{\max \tilde{y}}^{\max D} f\left(y\right) dy \simeq 0 \tag{6.24}$$

where $\tilde{y}$ denotes an empirically sampled observation.

Combining equations (6.21) and (6.23) yields the expression:

$$\ln O_k = \ln A_k + \varepsilon_{1,k} - \varepsilon_{2,k} \tag{6.25}$$

Substituting equation (6.22) into equation (6.25) gives:

$$\ln O_k = \ln\left(NW_k\right) + \ln f\left(y_k\right) + \varepsilon_{1,k} - \varepsilon_{2,k} \tag{6.26}$$

Finally, substituting equation (6.17) into equation (6.26) gives:

$$\ln O_k = \ln\left(NW_k\right) + \ln\left\{\exp\left[\sum_{j=1}^{J} \theta_j \psi_j\left(y_k\right) - \eta\right]\right\} + \varepsilon_{1,k} - \varepsilon_{2,k} \tag{6.27}$$

which can be written more simply as:

$$\ln\left(\frac{O_k}{NW_k}\right) = -\eta + \sum_{j=1}^{J} \theta_j \psi_j\left(y_k\right) + \varepsilon_{1,k} - \varepsilon_{2,k} \tag{6.28}$$

Equation (6.28) can be rearranged into the matrix linear regression form:

$$\ln\left(\frac{O_k}{NW_k}\right) = \Psi\theta + \varepsilon \tag{6.29}$$

where Ψ is a nonstochastic matrix composed by horizontally concatenating a unit vector with the column vectors $\psi_j(y_k)\ \forall j$, θ is amended to included the constant term and $\varepsilon = \varepsilon_{1,k} - \varepsilon_{2,k}$. Ordinary least squares parameter estimates of θ, in equation (6.29), are those of the LSFE. There is no need to apply numerical integration to calculate η (see equation (6.18)), as it is estimated as the negative of the constant term.

The LSFE, given by ordinary least squares estimates of equation (6.29), is inefficient. Efficient estimation of θ requires application of the covariance matrix Ω in the generalised least squares (GLS) framework, where:

$$\Omega \simeq \frac{1}{N}\begin{bmatrix} (O_1^{-1}N-1) & 1 & \dots & 1 \\ 1 & (O_2^{-1}N-1) & & \vdots \\ \vdots & & \ddots & 1 \\ 1 & \dots & 1 & (O_K^{-1}N-1) \end{bmatrix} \tag{6.30}$$

is derived in chapter 5. Given that Ω is a known, symmetric, positive definite matrix it is possible to derive the GLS estimates as:

$$\theta^* = \left(\Psi'\Omega^{-1}\Psi\right)^{-1}\Psi'\Omega^{-1}\ln\left(\frac{O_k}{NW_k}\right) \tag{6.31}$$

where:

$$var\left(\theta^*\right) = \left(\Psi'\Omega^{-1}\Psi\right)^{-1} \tag{6.32}$$

The parameter estimates, θ^*, are defined as belonging to the generalised least squares frequency estimator (GLSFE). For the remainder of this chapter, unconditional distribution parameter estimates are obtained using the GLSFE.

6.2.2 Maximum Likelihood Techniques

While the GLSFE, described in subsection 6.2.1, represents a convenient method for obtaining unconditional exponential family estimates from grouped data, it is not readily extended to conditional analysis. However, it is possible to obtain conditional exponential family parameter estimates by amending the standard (that is, nongrouped) log-likelihood function for grouped data. Lye and Martin (1994) define the conditional exponential family log-likelihood function derived from (6.17) as:

$$\ln L = \sum_{i=1}^{N}\left\{\sum_{j=1}^{J}\theta_{j,i}\left(z_{j,i}\right)\psi_j\left(y_i\right) - \eta_i\right\} \tag{6.33}$$

where N is the total number of individuals, $\theta_{j,i}$ is defined by (6.19) and:

$$\eta_i = \ln\int_D \exp\left[\sum_{j=1}^{J}\theta_{j,i}\left(z_{j,i}\right)\psi_j\left(y\right)\right]dy \tag{6.34}$$

Alterations to equation (6.33) are required when data are in grouped form. First, y is no longer a random variable, but restricted to taking class midpoint values, y_k, $k = 1, 2, \ldots, K$. It is convenient to consider the exogenous variable matrix z to be sorted such that unique exogenous variable characteristic sets (that is, identical rows of z) are contiguous. Unique rows in z constitute conditional distributions, where each distribution, indexed as $q = 1, 2, \ldots, Q$, is referenced by its own normalising constant. Therefore, the frequency data analogue of equation (6.33) is written:

$$\ln L = \sum_{q=1}^{Q} \sum_{k=1}^{K} O_{q,k} \left\{ \sum_{j=1}^{J} \theta_{j,q} \left(z_{j,q} \right) \psi_j \left(y_k \right) - \eta_q \right\} \tag{6.35}$$

where $O_{q,k}$ is the empirical frequency observed for the qth conditional distribution and kth class interval. Parameter estimates obtained by maximizing $\ln L$ in equation (6.35) will henceforth be referred to as frequency based maximum likelihood (FML) estimates.

6.3 Application to New Zealand Data

This section presents the unconditional and conditional distribution estimates based on grouped New Zealand income data. The form of the data sets is described in subsection 6.3.1. Although the demand and supply model presented in Section 6.1 yielded the generalised gamma distribution, variations in the model generated different special cases of the exponential family. For this reason, several alternatives are examined in subsection 6.3.2, using unconditional modelling. Subsection 6.3.3 reports conditional modelling estimates where the distributional parameters, θ, are expressed as functions of age.

6.3.1 New Zealand Income Distribution Data

The data were derived from a random sample of 2 per cent of IR3 and IR5 income tax returns filed for 1991. IR5 returns are filed by taxpayers whose income predominantly has tax withheld; that is, income from wages and salaries, taxable welfare benefits, NZ superannuation, interest and dividends. Individuals who have income from other sources or who have been paying provisional tax, file an IR3. The sample used here consists of those regarded as obtaining their income mainly from wages and salaries. If y and w denote taxable income and salary respectively, the condition $2w/3 \leq y \leq 4w/3$ was applied.

Individuals with income below $20,000 and who have tax withheld at source, who are not in receipt of income-related transfer payments or student loans, are not required to file. Such individuals may nevertheless choose to file, usually to receive a tax refund. Hence, people in the sample with incomes below $20,000 are not representative of the full tax-paying population with incomes below $20,000.

Taxable income is used for IR5 taxpayers, incorporating wage/salary income, interest, dividends and 'other' (e.g. illegal) income less tax agent fees for return preparation. For IR3 the 'income after expenses' definition is used, incorporating current losses, but not losses brought forward from prior years. This includes income from wages and salaries, interest, dividends, overseas, trusts, self employment, partnerships, shareholder salaries, rents, and 'other'. Private superannuation income is not included since this is not taxable in New Zealand.

Wages and salaries include all income from which PAYE is withheld at source. This includes welfare benefits, NZ superannuation, withholding payments (e.g. from shearing, domestic employment), redundancy payments, lump sum retiring allowances and taxable ACC payments. Welfare beneficiaries and NZ superannuitants cannot be separately identified, and their income is lumped together with any wage or salary income they may received from other jobs during the year. Age is calculated in years as at 31 March 1991. Gender is determined using a 'title' variable (Mr, Mrs, Brig., Rev. etc.). Where title was missing, sex was imputed assuming 50 per cent males. Where title was DR, 79 per cent males were assumed. Where title was numeric, the assumption was 92 per cent males (this situation seems to be restricted to army serial numbers). Data for the class intervals $(y \leq 0)$, $(180,000 < y \leq 200,000)$ and $(y > 200,000)$ are not used for estimation. For the male sample of $8,307$ observations this removes 42 individuals and for the female sample of $7,399$ individuals this removes 167 individuals.

For confidentiality reasons, the data are only available in grouped form. Frequency data for males and females are reported for classes containing the following income midpoints:

$$y_k = \begin{bmatrix} 2,500;\ 7,500;\ 12,500;\ 17,500;\ 22,500;\ 27,500;\ 32,500; \\ 37,500;\ 42,500;\ 47,500;\ 55,000;\ 65,000;\ 75,000; \\ 85,000;\ 95,000;\ 110,000;\ 130,000;\ 150,000;\ 170,000 \end{bmatrix} \tag{6.36}$$

where $k = 1, 2, \ldots, 19$. The first ten class intervals have a width of $5,000$, the next five class intervals have a width of $10,000$ and the last four class intervals have a width of $20,000$. Therefore, the class width vector is:

$$W_k = \begin{bmatrix} 5,000;\ 5,000;\ 5,000;\ 5,000;\ 5,000;\ 5,000;\ 5000; \\ 5,000;\ 5,000;\ 5,000;\ 10,000;\ 10,000;\ 10,000;\ 10,000; \\ 10,000;\ 20,000;\ 20,000;\ 20,000;\ 20,000 \end{bmatrix} \tag{6.37}$$

Table 6.1: New Zealand male grouped data: 1991

y_k: income	a_k: Age (midpoint)									
(midpoint)	22	27	32	37	42	47	52	57	62	67
2,500	16	17	9	5	3	6	4	2	1	3
7,500	117	43	46	39	27	33	36	44	117	147
12,500	105	61	55	43	26	25	23	22	141	160
17,500	113	79	61	50	40	27	19	29	38	26
22,500	198	148	132	99	95	72	58	67	30	12
27,500	164	218	207	151	126	107	87	67	31	9
32,500	91	183	193	156	131	116	80	75	41	3
37,500	46	131	165	152	133	123	84	56	23	4
42,500	22	88	137	144	129	97	67	35	30	8
47,500	10	45	92	116	103	74	56	39	11	4
55,000	3	23	97	108	101	92	64	35	15	1
65,000	2	15	30	38	44	56	33	21	9	0
75,000	2	4	9	20	25	15	21	12	5	0
85,000	1	1	7	10	10	9	6	3	3	0
95,000	0	0	4	10	5	8	2	0	1	1
110,000	0	1	0	6	4	8	3	3	2	0
130,000	0	1	0	1	3	5	5	0	1	0
150,000	0	0	3	0	2	0	0	0	0	0
170,000	0	0	0	3	2	1	1	1	0	0

Income data are subclassified by age groups where the following associated midpoints form the conditioning variable:

$$a_q = [22, 27, 32, 37, 42, 47, 52, 57, 62, 67] \tag{6.38}$$

where $q = 1, 2, \ldots, 10$.

Tables 6.1 and 6.2 show the available data for males and females respectively. The tables reveal that a relatively large number of cells for the female data set record zero observations. Moreover, comparison of Tables 6.1 and 6.2 demonstrate that females are highly represented by low income levels.

6.3.2 Unconditional Distribution Estimates

One method of determining the most appropriate statistical distribution for conditional analysis is to examine how well the density fits various unconditional empirical distributions. Therefore, this subsection compares the fit of alternative distributions from the exponential family to 1991 New Zealand

Table 6.2: New Zealand female grouped data: 1991

y_k: Income	a_k: Age (midpoint)									
(midpoint)	22	27	32	37	42	47	52	57	62	67
2,500	55	95	105	70	41	16	17	10	6	1
7,500	162	111	143	118	94	65	55	105	164	137
12,500	125	131	141	136	143	78	62	89	211	258
17,500	144	112	142	140	128	90	68	35	31	14
22,500	235	157	90	108	134	132	73	46	26	7
27,500	168	188	98	109	129	110	81	26	15	3
32,500	64	141	94	80	112	100	72	34	13	3
37,500	18	57	55	51	54	47	42	12	13	1
42,500	6	29	28	26	37	26	20	5	7	0
47,500	1	11	12	11	18	15	11	5	2	0
55,000	1	15	17	12	7	8	11	3	5	0
65,000	0	3	4	2	2	5	3	0	2	0
75,000	0	1	2	0	1	3	0	0	0	0
85,000	0	1	1	2	0	0	0	0	0	0
95,000	0	0	1	0	0	0	0	0	0	0
110,000	0	0	0	1	0	1	0	0	0	0
130,000	0	0	0	0	0	0	0	0	0	0
150,000	0	0	0	1	0	0	0	0	0	0
170,000	0	0	0	0	1	0	0	0	0	0

Table 6.3: Alternative distributions from the exponential family

Gen. gamma	$f(y) = \exp\left[\theta_1 \ln y + \theta_2 y + \theta_3 y^2 + \theta_4 y^3 - \eta\right]$
Gamma	$f(y) = \exp\left[\theta_1 \ln y + \theta_2 y - \eta\right]$
Gen. lognormal	$f(y) = \exp\left[\theta_1 y + \theta_2 \ln y + \theta_3 (\ln y)^2 + \theta_4 (\ln y)^3 - \eta\right]$
Lognormal	$f(y) = \exp\left[\theta_1 \ln y + \theta_2 (\ln y)^2 - \eta\right]$

unconditional income distributions classified by age group and sex. The fit criterion to be minimised is the chi-square statistic:

$$\chi^2 = \sum_k \frac{(f_{o,k} - f_{e,k})^2}{f_{e,k}} \tag{6.39}$$

where k indexes class intervals, $f_{o,k}$ is the unconditional empirical distribution's observed frequency in the kth class and $f_{e,k}$ is the expected frequency for the kth class. Expected frequencies are calculated as:

$$f_{e,k} = NW_k \exp(\Psi\theta^*)$$

where θ^* are obtained from the GLSFE given by equation (6.31) and Ψ is defined with reference to equation (6.29). Given that income is non-negative, only non-negative densities are considered as possible specifications for the most appropriate distribution. The distributions considered include the generalised gamma, gamma, generalised lognormal and lognormal. Functional forms for the distributions are given in Table 6.3.

Tables 6.1 and 6.2 apply the unconditional densities given in Table 6.3 to each age group of the male and female income data sets, respectively, as described in subsection 6.3.1. The generalised gamma density provides superior results in terms of smaller computed values of the chi-square statistic in nine of ten cases for males and six of ten cases for females. Only weak evidence supports application of the generalised lognormal density to model the 1991 New Zealand female income distribution.

To examine the support for the generalized gamma distribution observed in Tables 6.4 and 6.5, both male and female data are aggregated over all age groups and examined for the best fitting density. Results are reported in Table 6.6. The dominance of the generalised gamma distribution in modelling both the male and female empirical income distribution is again displayed. Therefore, empirical results suggest strong support for the economic model outlined in Section 6.1 that was used to derive the generalised gamma distribution. Hence, the following analysis is restricted to the generalised gamma distribution.

Table 6.4: χ^2 statistics: unconditional male income distributions

Age	Gen. Gamma	Gamma	Gen. Lognormal	Lognormal
22	72.54	111.94	106.22	243.68
27	44.82	223.37	128.38	509.08
32	50.23	217.06	138.32	411.87
37	66.86	226.04	138.91	373.31
42	46.07	156.74	93.37	344.60
47	44.99	111.48	87.48	272.30
52	65.34	96.00	88.98	171.44
57	53.14	79.07	72.50	94.18
62	113.63	197.34	58.55	94.01
67	45.56	408.67	62.96	124.98

Table 6.5: χ^2 statistics: unconditional female income distributions

Age	Gen. Gamma	Gamma	Gen. Lognormal	Lognormal
22	99.82	226.18	53.67	404.48
27	49.16	218.34	83.39	447.93
32	22.22	48.58	28.24	136.92
37	18.87	73.87	32.64	122.65
42	15.79	183.71	61.21	178.51
47	25.59	65.03	53.62	162.76
52	12.77	39.86	7.97	92.59
57	21.69	54.52	19.73	33.57
62	68.82	207.93	71.37	103.73
67	52.50	124.05	21.56	103.80

Table 6.6: χ^2 statistics: income distributions aggregated over age

Sex	Gen. Gamma	Gamma	Gen. Lognormal	Lognormal
Male	493.71	828.40	849.83	1593.50
Female	227.86	858.68	826.33	1080.28

6.3.3 Conditional Model Specification

This subsection extends the unconditional generalised gamma distribution by allowing the distributional parameters, $\theta_{j,q}$, to be functions of age, a_q. Adopting the notation of equation (6.19) gives:

$$\theta_{j,q} = g_j (a_q) \tag{6.40}$$

Selecting the functional forms, $g_j (.)$, is a nontrivial task. One approach is to express each θ_j as a polynomial function of age, and successively eliminate statistically insignificant terms. However, given the considerable data limitations present, this approach is not followed. Alternatively, the structural parameters $(\alpha, \beta, \gamma, \delta)$ can be expressed as polynomial functions of age and substituted into equations (6.12) to (6.15). This approach is not adopted given the high degree of nonlinearity introduced and the lack of *a priori* motivation.

The methodology adopted involves examining various unconditional distributions in order to determine the most appropriate conditional specification. Because age, a_q, is the only conditioning variable considered, an unconditional generalised gamma distribution was estimated using the GLSFE for each age group. Plots of $\left(\hat{\theta}_{1,q}, \hat{\theta}_{2,q}, \hat{\theta}_{3,q}, \hat{\theta}_{4,q}\right)$ against a_q are presented in Figure 6.1 for males. By displaying how the $\hat{\theta}_{j,q}$ vary with a_q, the plots provide insight into the appropriate form of $g_j (.)$ in equation (6.40).

For the male data, an approximately quadratic relationship is demonstrated between $\left(\hat{\theta}_{1,q}, \hat{\theta}_{2,q}\right)$ and a_q. Although a similarly quadratic relationship appears between $\hat{\theta}_{3,q}$ and a_q, subsequent estimation results favoured inclusion of the second-order term only. The term θ_4 is restricted to be negative during estimation, thus ensuring that the density does not become explosive as $y \rightarrow \infty$. Therefore, the following conditional stationary distribution specification is adopted to model the 1991 New Zealand male income distribution:

$$f_s (y) = \exp \left[\theta_{1,q} \ln y + \theta_{2,q} y + \theta_{3,q} y^2 + \theta_{4,q} y^3 - \eta_q\right] \tag{6.41}$$

where:

$$\begin{aligned} \theta_{1,q} &= \gamma_1 + \gamma_2 a_q + \gamma_3 a_q^2 && (6.42) \\ \theta_{2,q} &= \gamma_4 + \gamma_5 a_q + \gamma_6 a_q^2 && (6.43) \\ \theta_{3,q} &= \gamma_7 + \gamma_8 a_q^2 && (6.44) \\ \theta_{4,q} &= - |\gamma_9| && (6.45) \end{aligned}$$

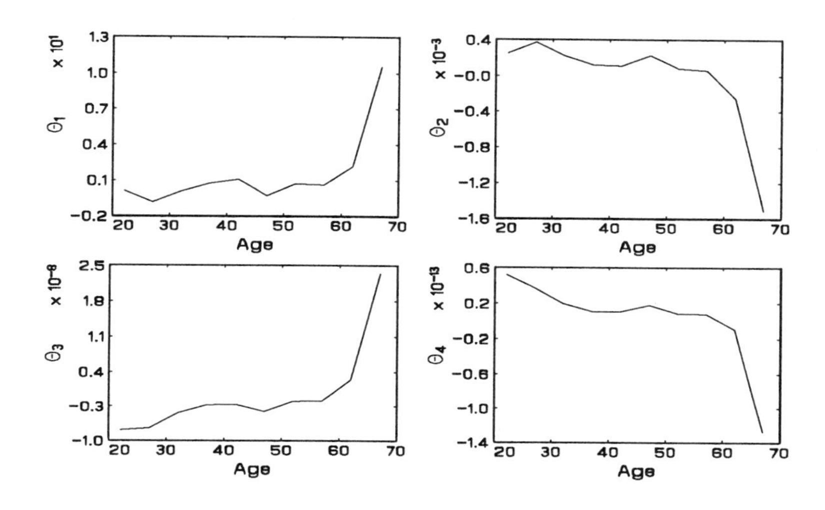

Figure 6.1: Male unconditional distribution parameters

Although the same methodology was applied to determine a conditional specification for the female data, it was not obvious from the resulting diagrams what form the distribution should take. Hence, the male specification was also adopted for female data. During estimation y_k were divided by (1000×52), thus converting income into a thousand dollar per week measure and a_q^2 is divided by 1000. This scaling improved the convergence behaviour of the GAUSS maximum likelihood algorithms applied to equation (6.35).

FML, as described in subsection 6.2.2, is used to obtain parameter estimates of γ and associated standard errors for 1991 male and female data as presented in Tables 6.1 and 6.2. Estimation results are reported in Table 6.7. Except for the male data estimates of $\hat{\gamma}_1$ and $\hat{\gamma}_9$, all parameter estimates in Table 6.7 are statistically significant. However, the properties of standard errors for FML are unknown, so caution with interpretation is recommended.

Parameter estimates in Table 6.7 can be used to construct probability density surfaces for males and females that have age and income as x-y axes. Both fitted densities capture the features of their corresponding empirical distributions, namely, high peaks at young and elderly age groups, with a relatively higher peak observed for the elderly and more dispersed distributions corresponding to middle-aged people. The female fitted density surface appears relatively more skewed than the male fitted density surface, again

Table 6.7: Generalized gamma conditional distribution: FML estimates

	Male		Female	
Parameter	Estimate	Std. err.	Estimate	Std. err.
γ_1	0.2587	0.4847	3.4439	0.3967
γ_2	0.1209	0.0261	−0.2011	0.0214
γ_3	−1.5363	0.3030	3.2483	0.2738
γ_4	−15.2367	1.0116	−14.6234	1.3666
γ_5	0.6349	0.0557	1.1446	0.0742
γ_6	−8.1909	0.7096	−19.4784	1.0388
γ_7	−2.0907	0.2463	−7.2293	0.4074
γ_8	0.7015	0.0589	3.0154	0.2172
γ_9	−0.0002	0.0028	−0.4817	0.1154
$\ln L$	−6.3057		7.4479	

consistent with empirical observation.

6.4 Conclusions

This chapter examines grouped frequency distributions of taxable income for New Zealand males and females. The distributions were modelled from labour demand and supply analysis applied to a continuous time error-correction stochastic process that generates a stationary income distribution from the exponential family. Although the generalised gamma distribution was derived, variations in the labour demand and supply model give rise to different special cases of the family. The stationary distribution was extended to a conditional form by expressing the distributional parameters as functions of age.

The generalised gamma distribution was adopted for conditional modelling because, of the distributions examined, it provided the best fit to the income distributions of particular age groups, thus supporting the initial supply and demand model used to derive the distribution. A least squares frequency-based estimator was shown to be a convenient method of estimating these unconditional distributions. Following further examination of the unconditional distributions in each age group, three of the four parameters of the generalised gamma distribution were expressed as quadratic functions of age. The resulting nine parameters were then estimated using maximum likelihood (modified for grouped data). The estimated model captured the major empirical features of the changing distribution of income with age, namely, high peaks at young and elderly age groups, with a relatively higher peak observed for the elderly and more dispersed distributions corresponding to middle-aged people.

It is suggested that the flexibility of the model offers much potential for further income distribution modelling; for example, it could be applied to particular occupations. Furthermore, the approach could easily be adapted to the analysis of the changing aggregate distribution over time, thereby allowing the influence of macroeconomic factors on the income distribution to be estimated.

Chapter 7

Count Data and Discrete Distributions

David Dickson and Vance L. Martin

This chapter introduces a broad class of discrete distributions for modelling the unconditional distribution of count data. The class is motivated by a discretisation of the generalised Pearson differential equation investigated by Cobb *et al.* (1983) and Lye and Martin (1993a). The derived distribution is referred to as a generalised exponential discrete distribution. Using examples relating to claim numbers under insurance policies it is shown that this flexible class produces fitted distributions which compare favourably with results of fitting a range of existing discrete distributions.

In Section 7.2 alternative classes of distributions used in the actuarial literature to model claim numbers data are discussed. The generalised exponential distribution is presented in Section 7.3 as an alternative to these existing classes of discrete distributions, while estimation issues are discussed in Section 7.4. Section 7.5 contains the empirical results where a range of distributions are fitted to insurance claims data, and Section 7.6 provides a discussion of estimating aggregate claims distributions when the claim number distribution is a generalised exponential distribution.

7.1 Families of Discrete Distributions

In this section three types of discrete distributions are summarised which are used to model the number of claims arising from an insurance policy. Let N be a discrete random variable with probability function $p_n = \Pr(N = n)$ for $n = 0, 1, 2, \ldots$.

A first class of discrete distributions is the polynomial class defined by:

$$p_n = \frac{\alpha(n)}{\beta(n)} p_{n-1} \quad \text{for } n = 1, 2, 3, ... \tag{7.1}$$

where $\alpha(n) = \sum_{i=0}^{K} \alpha_i n^i$ and $\beta(n) = \sum_{i=0}^{M} \beta_i n^i$. Contained within this class are a number of important distributions. In particular, when:

$$p_n = \frac{\alpha_0 + \alpha_1 n}{n} p_{n-1} \quad \text{for } n = 1, 2, 3, ... \tag{7.2}$$

the distributions which arise are Poisson, binomial and negative binomial; see, for example, Panjer and Willmot (1992, p.195). Examples of distributions satisfying (7.1) when $K = M = 2$ are given by Hesselager (1994).

A second class of discrete distributions is the class of mixed Poisson distributions. In this class the probability function of N is defined by:

$$p_n = \int_{\lambda} e^{-\lambda} \frac{\lambda^n}{n!} f(\lambda) d\lambda$$

where $f(\lambda)$ is a density function, referred to as the mixing distribution. In the examples of Section 7.5 the situation where $f(\lambda)$ has a gamma distribution shifted by $\gamma \geq 0$ is considered. In the case when $\gamma > 0$, the resulting distribution is the Delaporte distribution; see, for example, Panjer and Willmot (1992, p.289); and when $\gamma = 0$ the resulting distribution is the negative binomial distribution. An extensive discussion of recursive evaluation of mixed Poisson distributions is given by Willmot (1993).

A third class of discrete distributions is the class of compound distributions. In this case the model for N is:

$$N = \sum_{i=1}^{R} Y_i \tag{7.3}$$

where R is a discrete random variable and $\{Y_i\}_{i=1}^{\infty}$ is a sequence of independent and identically distributed non-negative valued discrete random variables. This sequence is independent of R. An example of a situation in which such a model might be applied would be if an employer had an insurance policy providing compensation for employees injured at work. In this context, N represents the number of claims, R represents the number of accidents, and Y_i is the number of claims from the ith accident. In this model the distribution of R is referred to as the primary distribution, and the distribution of Y_i is referred to as the secondary distribution. A full discussion of compound counting distributions is given in Panjer and Willmot (1992).

These three classes represent the types of distribution most commonly used to model insurance claim counts data.

The main application of such distributions is in modelling the aggregate claim amount arising from a policy, or a portfolio of policies, over a fixed period. The aggregate claim amount is modelled as:

$$S = \sum_{i=1}^{N} X_i \tag{7.4}$$

where S denotes the aggregate claim amount and X_i denotes the amount of the ith claim. It is assumed that $\{X_i\}_{i=1}^{\infty}$ is a sequence of independent and identically distributed random variables, independent of N. In the particular case when X_i is a discrete random variable with probability function $\{f_j\}_{j=1}^{\infty}$ the probability function of S, denoted by g_x for $x = 0, 1, 2, ...$, is given by $g_0 = p_0$ (since $S = 0$ only if $N = 0$) and:

$$g_x = \sum_{n=1}^{\infty} p_n f_x^{n*} \quad \text{for } x = 1, 2, 3, .. \tag{7.5}$$

where $f_x^{n*} = \Pr(\sum_{i=1}^{n} X_i = x)$ is the n-fold convolution of the probability function f_x with itself.

In general, the problem with evaluating (7.5) is that it is time consuming to compute convolutions. However, efficient calculation of g_x through a recursion formula is possible in certain circumstances. If the distribution of N belongs to the polynomial class (7.1) then a recursive calculation is possible; see Hesselager (1994). In particular, if N belongs to the class given by (7.2) this yields the celebrated Panjer recursion formula; see Panjer and Willmot (1992). For example, if N has a Poisson distribution with parameter λ, the simple recursion formula is:

$$g_x = \frac{\lambda}{x} \sum_{j=1}^{x} j f_j g_{x-j} \quad \text{for } x = 1, 2, 3, .. \tag{7.6}$$

with $g_0 = p_0 = e^{-\lambda}$. Such recursion formulae do not exist in general if N is a member of the other two classes of distribution described above. However, for certain mixed Poisson distributions, recursion formulae for g_x exist; see Willmot (1993) and Sundt (1992).

Thus in modelling insurance claim numbers data, one important consideration is the application of the resulting distribution. If it is desired to calculate the probability function g_x it is tempting to fit a distribution for which a recursion formula for g_x exists, as direct calculation of g_x from formula (7.5) is not a realistic option. However, if the distribution which best fits data does not result in a recursive calculation for g_x then the distribution of S can always be approximated provided it is possible to calculate moments of the distributions of N and X_i. Details of such methods can be found in Beard *et al.* (1984, ch. 3).

7.2 Generalised Exponential Distributions

In this section the generalised exponential class of distributions is introduced. This class is derived from a discretisation of the generalised Pearson differential equation investigated by Cobb *et al.* (1983) and Lye and Martin (1993a). If a continuous random variable has density function $p(x)$, the pertinent differential equation is:

$$\frac{d}{dx} \ln p(x) = \frac{a(x)}{b(x)} \tag{7.7}$$

where $a(x)$ and $b(x)$ are some functions. A discretisation of (7.7) of length h is:

$$\frac{\ln p(x) - \ln p(x-h)}{h} = \frac{a(x)}{b(x)} \tag{7.8}$$

where the derivative in (7.7) is approximated by a simple numerical backward derivative. In the case of discrete data where the step length is $h = 1$, (7.8) is written as:

$$\ln p(x) - \ln p(x-1) = \frac{a(x)}{b(x)}$$

or, alternatively:

$$p_n = \exp\left(\frac{a(n)}{b(n)}\right) p_{n-1} \tag{7.9}$$

A particular class of this distribution arises when $a(n)$ and $b(n)$ are polynomials, giving:

$$p_n = \exp\left(\frac{\sum_{i=0}^{K} a_i n^i}{\sum_{i=0}^{M} b_i n^i}\right) p_{n-1} \tag{7.10}$$

For certain types of data sets, it may be difficult to model the tail behaviour. In particular, the empirical distribution may exhibit a relatively fatter right-hand tail than any of the standard distributions discussed in the previous section. The problem in applying (7.10) is trading off flexibility for parsimony: by allowing K and M to increase, greater flexibility is achieved at the cost of the inclusion of additional parameters.

One way to circumvent this problem is to approximate the polynomial expressions by a nonlinear function or a set of functions. The form proposed is:

$$\begin{aligned} a(n) &= a_0 + a_1(n-1)^K \\ b(n) &= n^M \end{aligned} \tag{7.11}$$

giving the generalised exponential distribution, denoted by $GE(K, M)$, as:

$$p_n = \exp\left(\frac{a_0 + a_1(n-1)^K}{n^M}\right) p_{n-1} \quad \text{for } n = 1, 2, 3, ... \tag{7.12}$$

This distribution has four parameters: a_0, a_1, K, M. While other parameterizations are clearly possible, this specification should be flexible enough to capture a wide range of distributional shapes. Specifications that restrict K and M to integer values, say 1 or 2, the exponent of (7.12) contains a ratio of finite polynomials. For these cases the generalised exponential distribution mimics the standard distributions contained in (7.2). For noninteger values of K and M, then (7.12) contains a ratio of infinite polynomials.

For data sets with only a small range of observations, for example $n = 0, 1, 2, ..., 5$, it may be preferable to fit a distribution with fewer parameters, say with K and M fixed. In the examples in Section 7.5 methods of specifying M and K to reduce the number of estimated parameters are discussed. In particular, as is shown, the fitted four parameter generalised exponential distributions do not necessarily perform better than those with a smaller number of estimated parameters.

7.3 Estimation Procedures

For estimation purposes, (7.12) is written as:

$$p_n = \gamma(n) p_{n-1} \tag{7.13}$$

where:

$$\gamma(n) = \exp\left(\frac{a_0 + a_1(n-1)^K}{n^M}\right)$$

To ensure that a proper distribution is obtained, it is required that

$$\lim_{n \to \infty} \gamma(n) < 1$$

so that the tail of the distribution goes to zero as $n \to \infty$. When $K \geq M$ this occurs provided that both a_0 and a_1 are negative. For this case the model can be estimated as follows. By repeated application of (7.13):

$$p_n = \prod_{i=1}^{n} \gamma(n) p_0 = \pi(n) p_0 \tag{7.14}$$

and so:

$$p_0 = \frac{1}{1 + \sum_{j=1}^{\infty} \pi(j)} \tag{7.15}$$

Since $\lim_{n\to\infty}\gamma(n) < 1$, this implies that $\lim_{n\to\infty}\pi(n) = 0$. In order to calculate p_0 in (7.15) the summation in the denominator is truncated at 100. Although this choice is somewhat arbitrary, it is found that in the empirical applications the parameter estimates are insensitive to alternative, higher truncations.

In the examples in the next section, parameter values are estimated by maximum likelihood methods. In particular, if n_j, $j = 0, 1, 2, ..., J$, are the observed frequencies associated with the probabilities p_j, then the log-likelihood function is given by:

$$\ln L = \sum_{j=0}^{J} n_j \ln p_j \tag{7.16}$$

The log-likelihood function is maximised using standard gradient algorithms. Given that the log-likelihood function is in general a nonlinear function of the parameters of the model, it is convenient to use numerical derivatives. The gradient algorithm used in the computations is based on the GAUSS applications module OPTMUM. In calculating values of $\ln L$ it is necessary first to calculate p_0 given a set of parameter values from (7.15), then evaluate p_n values from (7.14).

In the case where the parameter M is estimated, or if the model is estimated subject to the restriction $M > K$, it is necessary to alter the algorithm. The reason for this is that in this case $\lim_{n\to\infty}\gamma(n) = 1$, so that the recursion formula does not produce probability estimates that decrease as n goes to infinity. A pragmatic approach, although subjective, is to assume that the recursion formula (7.13) holds for $n = 1, 2, 3, ..., n_0$, with $p_n = 0$ for $n > n_0$. The choice of the truncation point n_0, depends on the problem being considered. In the examples considered in this chapter concerning the number of claims arising in a year under an automobile insurance policy, a sensible approach is to assume that a policyholder can make at most 20 claims in a year. This choice for n_0, while subjective, is consistent with both the observed data and *a priori* expectations about claim numbers under such policies.

7.4 Numerical Illustrations

In this section the properties of fitting the generalised exponential distribution to insurance claim numbers data are compared with a range of existing distributions. The first data set comes from Thyrion (1961) and relates to the number of claims per policy under automobile insurance policies. This data set has been analysed by Ruohonen (1988). The distributions fitted to these data are as follows:

(i) Negative binomial distribution, denoted by NB in the tables:

$$p_n = \frac{\Gamma(n+\alpha)}{\Gamma(\alpha)n!}\left(\frac{\beta}{1+\beta}\right)^{\alpha}\left(\frac{1}{1+\beta}\right)^{n} \quad \text{for } n = 0, 1, 2, ... \tag{7.17}$$

(ii) Delaporte distribution, denoted by Del in the tables:

$$p_n = \sum_{k=0}^{n} \frac{\Gamma(\alpha+k)}{\Gamma(\alpha)k!}\left(\frac{\beta}{1+\beta}\right)^{\alpha}\left(\frac{1}{1+\beta}\right)^{k} \frac{\gamma^{n-k}e^{-\gamma}}{(n-k)!} \quad \text{for } n = 0, 1, 2, ... \tag{7.18}$$

(iii) Modified extended truncated negative binomial distribution, denoted by METNB in the tables:

$$p_n = (1-\pi)\frac{-\Gamma(n+\alpha)}{\Gamma(\alpha)n!}\frac{p^n}{1-(1-p)^{-\alpha}} \quad \text{for } n = 1, 2, 3, ... \tag{7.19}$$

with $p_0 = \pi$.

(iv) Polya-Aeppli distribution, denoted by P-A in the tables. This is a compound distribution where the primary distribution is Poisson with parameter λ and the secondary distribution is geometric with probability function:

$$\Pr(Y_i = j) = \frac{\beta^{j-1}}{(1+\beta)^j} \quad \text{for } j = 1, 2, 3, ... \tag{7.20}$$

where Y_i is as in (7.3).

(v) Generalised exponential distributions, referred to as $GE(K, M)$ in the tables.

Table 7.1 shows the observed numbers of claims per policy, denoted O_k, as well as the expected numbers under the different models. The frequencies are rounded to the nearest integer, which may mean that the sum of the expected numbers does not match the total of the observed numbers in some cases. Also given are the value of $-\ln L$, the chi-squared test statistic, the number of degrees of freedom and the p-value for the test statistic.[1] Table 7.2 gives the estimated parameter values for the $GE(K, M)$ distributions.

[1]In computing the number of degrees of freedom, the expected frequency in each cell is constrained to be at least equal to 5.

Table 7.1: Comparison of observed and expected claims: data set I

k	O_k	NB	Del	METNB	$GE(1,1)$	$GE(2,M)$	$GE(K,M)$
0	7840	7847	7837	7840	7849	7840	7840
1	1317	1288	1326	1320	1286	1317	1318
2	239	257	223	225	256	236	234
3	42	54	53	54	54	48	48
4	14	12	15	15	12	13	13
5	4	3	5	4	3	4	4
6	4	1	2	1	1	2	2
7	1	0	1	0	0	1	1
$-\ln L$		5348	5343	5344	5348	5342	5342
χ^2		8.817	4.312	4.868	8.523	1.715	1.733
DF		2	2	2	2	2	1
p-value		0.012	0.116	0.088	0.014	0.439	0.188

Table 7.2: Estimated parameter values: data set I

	$GE(1,1)$	$GE(2,M)$	$GE(K,M)$
$\widehat{a}_0$	−1.809	−1.783	−1.783
$\widehat{a}_1$	1.418	−16.596	−14.366
$\widehat{K}$			3.226
$\widehat{M}$		3.416	1.884

Of the distributions fitted to these data, the $GE(2,M)$ model gives the best fit, although it is clear that the hypotheses that the data come from the Delaporte distribution or from the modified extended truncated negative binomial distribution would not be rejected. One advantage of fitting the Delaporte distribution to these data is that the aggregate claims distribution can be calculated from a recursion formula in this case.

In the second example the data are taken from Simon (1961), and as with the first example, also involve automobile insurance. Table 7.3 shows the observed number of claims per policy, denoted O_k, as well as the expected numbers under the three best-fitting standard distributions listed above, and under three generalised exponential distributions. Table 7.4 shows the estimated parameter values for the $GE(K,M)$ distributions.

Inspection of Table 7.3 shows that a number of distributions could reasonably be fitted to these data. In particular, all the generalised exponential

Table 7.3: Comparison of observed and expected claims: data set II

k	O_k	Del	METNB	P-A	$GE(2,2)$	$GE(2,M)$	$GE(K,M)$
0	99	100	99	99	99	98	99
1	65	67	69	71	69	72	66
2	57	52	51	50	51	51	56
3	35	33	33	33	34	32	35
4	20	20	20	20	21	19	19
5	10	12	12	12	12	11	10
6	4	6	6	7	7	6	5
7	0	4	4	4	4	4	3
8	3	2	2	2	2	2	2
9	4	1	1	1	1	1	1
10	0	1	1	0	0	1	1
11	1	0	0	0	0	0	0
$-\ln L$		528	528	528	528	528	527
χ^2		2.017	2.255	2.877	2.434	2.446	0.678
DF		4	4	5	5	4	3
p-value		0.733	0.689	0.719	0.786	0.653	0.878

Table 7.4: Estimated parameter values: data set II

	$GE(2,2)$	$GE(2,M)$	$GE(K,M)$
$\widehat{a}_0$	−0.365	−0.312	−0.404
$\widehat{a}_1$	−0.837	−1.291	−18.169
$\widehat{K}$			5.432
$\widehat{M}$		2.258	6.762

models provide good fits, with the p-value for the test statistic under the generalised exponential distribution with four fitted parameters being particularly high.

7.5 Aggregate Claims Distributions

In this section the question is addressed of calculating aggregate claims distributions when the claim number distribution is generalised exponential. The most obvious way to do this is to approximate the aggregate claims distribution by a distribution which has the same first few moments as the aggregate claims distribution. Thus, the moments of the $GE(K, M)$ distribution are required. In the case where $K \geq M$ these are easily obtained from the formula:

$$E(N^r) = \sum_{n=0}^{n_0} n^r p_n \quad \text{for } r = 1, 2, 3, ... \qquad (7.21)$$

where n_0 is a suitably large value. A natural choice for the value of n_0 is the point at which the summation in the denominator of (7.15) is truncated in the estimation procedure.

When $K < M$, the moments can be calculated from (7.21), except that the upper limit of summation is n_0. In estimating the $GE(2, M)$ models in the previous section $n_0 = 20$. These models are also estimated with $n_0 = 15$ and $n_0 = 25$. Changing the value of n_0 makes no real difference to the expected frequencies in Tables 7.1 and 7.3, but does have some effect on the moments, as shown in Tables 7.5 and 7.6.

Inspection of Tables 7.5 and 7.6 show that, in these examples at least, both the mean and variance of the fitted $GE(2, M)$ distribution vary little over a range of intuitively sensible values for n_0. In these examples it would certainly be unrealistic to set the value of n_0 any higher than 25. These tables illustrate that fitting these $GE(2, M)$ models produces acceptable results.

Using the moments in Tables 7.5 and 7.6, the aggregate claims distribution using one of the moment-based approximation methods described in

Table 7.5: Moments of estimated GE(2,M) models: data set I

n_0	$E(N)$	$V(N)$	$E([N-E(N)]^3)$
15	0.2145	0.2916	0.5897
20	0.2145	0.2924	0.6104
25	0.2143	0.2926	0.6258

Table 7.6: Moments of estimated GE(2,M) models: data set II

n_0	$E(N)$	$V(N)$	$E([N-E(N)]^3)$
15	1.702	3.626	11.41
20	1.704	3.641	11.74
25	1.704	3.643	11.80

Beard *et al.* (1984) can be approximated. By assuming that claim numbers from individual policies are independent, the second and third central moments of the number of claims from the portfolio are easily calculated as the sum of these quantities for individual policies. If the expected number of claims from the portfolio is reasonably large, moment-based approximations produce reasonably accurate results, particularly if upper percentiles of the aggregate claims distribution are desired.

For data set I, the fitted Delaporte distribution in (7.18) which has the highest p-value after the $GE(2,M)$ distribution has the following estimated parameters:

$$\widetilde{\alpha} = 0.2006, \quad \widetilde{\beta} = 1.6665, \quad \widetilde{\gamma} = 0.0940$$

These give the following estimates of the mean and the second and third central moments of the distribution as 0.2144, 0.2866 and 0.5178 respectively. This distribution exhibits virtually the same mean as the $GE(2,M)$ distribution but has smaller second and third moments. Thus the $GE(2,M)$ distribution yields an aggregate claims distribution which exhibits greater variability and skewness than if the counting distribution is Delaporte. Sample calculations using moment-based approximations show that the percentiles of aggregate claims distributions using each of these counting distributions are very similar, with the values being slightly higher when the counting distribution is $GE(2,M)$.

These results show that it is possible to estimate moments and approximate percentiles of the aggregate claims distribution when the claim number distribution is $GE(K,M)$. Whilst the estimation of moments is subjective when $M > K$, Tables 7.5 and 7.6 indicate that an appropriate choice of n_0 leads to acceptable values. The calculation of percentiles is not as precise as it could be using a recursion formula to calculate the aggregate claims distribution. Nevertheless, moment-based approximations are useful in practice and can capture the difference in variability between $GE(K, M)$ and other counting distributions.

7.6 Conclusions

This chapter investigated a new class of discrete distributions. This class was derived as a discrete approximation to the continuous generalised exponential family. Estimation issues were also investigated. The generalised exponential class of discrete distributions was fitted to two insurance count data sets and found to be superior in terms of goodness of fit to a range of existing discrete distributions.

The application of the fitted distribution raised two issues requiring further study. The first concerns the choice of the truncation parameter n_0 when $K > M$. The second concerns devising an efficient algorithm for evaluating the aggregate claims distribution when the counting distribution is generalised exponential.

Part III

Time Series Applications

Chapter 8

A Model of the Real Exchange Rate

John Creedy, Jenny N. Lye and Vance L. Martin

Most models of the exchange rate are linear, although evidence suggests that linear models are not consistent with the data; see Meese and Rogoff (1983), Meese and Rose (1991) and Diebold and Nason (1990). The underlying assumption of nonlinear models is that while the set of market fundamental variables commonly used in empirical exchange rate models is appropriate, the relationship between the exchange rate and market fundamentals is not linear. The approach adopted is to identify the source of the nonlinearity through a range of nonlinear empirical techniques. Recent examples include the random coefficient model of Schinasi and Swamy (1989), the use of nonparametric estimation procedures by Meese and Rose (1991) and Chinn (1991); the stochastic segmented trends model of Engel and Hamilton (1990); and the adoption of an ARCH specification of the variance by Diebold (1988) and Domowitz and Hakkio (1985). The first three examples correspond to specifying nonlinear models of the mean, while the last corresponds to a nonlinear model of the variance.

This chapter presents an alternative nonlinear model of the exchange rate. The main differences from existing nonlinear approaches are that the specification is parametric, and that the third and fourth moments are considered in addition to the mean and the variance. The approach is based on the work of Creedy and Martin (1993, 1994b, c) which consists of building a nonlinear model by solving for the real exchange rate distribution from a flexible continuous-time error correction model. The derived distribution belongs to the generalised exponential class, introduced in chapter 1. A property of this distribution is that it encompasses the normal distribution as well as being able to capture skewness, kurtosis and even multimodality. This latter property is found to be important when applying the framework to the real US/UK exchange rate over the period 1973 to 1990, since the large swings observed in the real exchange rate correspond to periods when the distribution is bimodal.

In Section 8.1 a nonlinear model of the real exchange rate is derived. Estimation and prediction conventions are discussed in Section 8.2. The framework is applied in Section 8.3 to the US/UK real exchange rate using monthly data for the period from 1973:3 to 1990:5; these data were used by Baillie and Pecchenino (1991). For comparative purposes, the forecasting properties of the nonlinear model are compared with the predictions based on linear specifications which include alternative random walk specifications and speculative efficiency, and the nonlinear exchange rate model of Engel and Hamilton (1990) which is based on a stochastic segmented trends model.

8.1 The Model

A nonlinear model of the exchange rate should be sufficiently flexible, but must have sufficient restrictions placed on it so as to limit the number of choices from a very large set of nonlinear functional forms. Consider the general reduced-form relationship between the real exchange rate s, and a set of market fundamental variables contained in the vector x:

$$\mu(s, x) = 0 \tag{8.1}$$

In the empirical application below, this relationship occurs at each point in time. However, it is convenient in the derivation of the model to suppress the time subscript.

Economic models of exchange suggest that it is appropriate to include higher-order terms in s at least up to order three; see the class of models examined in Creedy and Martin (1993). If (8.1) is characterised by a cubic in s, it can either display at least one root which corresponds to a single equilibrium, or at most three roots which occur when there are multiple equilibria. A parameterisation to capture these properties is obtained, following Creedy and Martin (1994b, c), by writing (8.1) as:

$$\mu(s, x) = \alpha_0(x) + \alpha_1(x)s + \alpha_2(x)s^2 + \alpha_3(x)s^3 \tag{8.2}$$

where $\alpha_j(x)$ represent general functions of the market fundamental variables x.

8.1.1 The Error Correction Representation

Stochastics can be added to the model using the following error correction model:

$$ds = -\mu(s,x)dt + \sigma(s,x)dW \tag{8.3}$$

where $\mu(s,x)$ represents the (instantaneous) mean which is given by (8.2), $\sigma^2(s,x)$ is the (instantaneous) variance which in general is a function of s and the market fundamental variables in x, and W is a Wiener process with the property that dW is distributed as $N(0,dt)$. The stochastics of the model are captured by the term dW which represents the continuous time analogue of the normal distribution.

Equation (8.3) is an Ito process which shows that the real exchange rate adjusts continuously from its mean as a result of stochastic shocks, dW; see Kamien and Schwartz (1981). The negative sign means that a positive shock, that is $dW > 0$, leads to an increase in the real exchange rate but that over time the real exchange rate, on average, error corrects to the original equilibrium position.

8.1.2 Distributional Dynamics

Associated with equation (8.3) is the density function of s, denoted by $f(s)$. Consider a shock to the system and assume that adjustment to equilibrium is not instantaneous, so that over time the process moves through a sequence of temporary equilibria. Associated with each temporary equilibrium is the transitional distribution of s. The dynamics of the process are summarised by the Kolmogorov forward equation (see Cox and Miller, 1984, p.208):

$$\frac{\partial f}{\partial t} = \frac{\partial}{\partial s}(\mu f) + \frac{1}{2}\frac{\partial^2}{\partial s^2}\left(\sigma^2 f\right) \tag{8.4}$$

This partial differential equation defines the transitional distribution of the real exchange rate at each point in time. For a given setting of the market fundamentals x, the sequence of real exchange rate transitional distributions converges to a stationary distribution. If the market fundamentals remain constant the observed real exchange rates represent realisations of drawings from the same stationary distribution. When the market fundamentals do change over time, associated with each market fundamental setting at a particular point in time is a new sequence of transitional distributions which converges to a new stationary distribution. If prices are flexible, the speed of convergence to the stationary distribution is fast. For a chosen sampling frame, the observed real exchange rate at a particular point in time represents a drawing from the stationary distribution at that point in time.

Except for very simple expressions for both the mean and the variance, no analytical solution of (8.4) exists. However, as discussed in chapter 1, the stationary distribution, f^*, can be derived. This is given by:

$$f^*(s) = \exp\left[-\int_0^s \left(\frac{2\mu(w) + d\sigma^2(w)/dw}{\sigma^2(w)}\right) dw - \eta^*\right] \tag{8.5}$$

where $\eta^*$is the normalising constant, determined by the boundary conditions of (8.4).

In specifying the variance of the real exchange rate, it is often assumed that exchange rate volatility increases with the level of the real exchange rate (Malliaris and Brock, 1982). Thus, an appropriate specification for $\sigma^2(s)$ is:

$$\sigma^2(s) = \gamma s \tag{8.6}$$

To obtain the stationary distribution, substitute (8.2) and (8.6) in (8.5) and solve to get:

$$f_G^* = \exp\left[\pi_1(x)\ln(s) + \pi_2(x)s + \pi_3(x)s^2 + \pi_4(x)s^3 - \eta^G\right], 0 < s < \infty \tag{8.7}$$

where η^G is the normalising constant, and

$$\begin{aligned}
\pi_1 &= -(2\alpha_0(x) + \gamma)/\gamma \\
\pi_2 &= -2\alpha_1(x)/\gamma \\
\pi_3 &= -\alpha_2(x)/\gamma \\
\pi_4 &= -2\alpha_3(x)/3\gamma
\end{aligned}$$

This distribution is a generalised gamma distribution and contains as a special case the standard gamma distribution ($\pi_3 = \pi_4 = 0$).

In most empirical work on exchange rates, the logarithm of the exchange rate is used. The distribution of $e = \ln(s)$ is obtained by using the transformation technique to write the distribution in (8.7) in terms of e. Expanding the exponential terms in Taylor series expansions up to order four around the parity level of $e = 0$, that is $s = 1$, and rearranging the terms gives

$$f_G^* = \exp\left[\delta_1(x)e + \delta_2(x)e^2 + \delta_3(x)e^3 + \delta_4(x)e^4 - \eta^N\right], -\infty < e < \infty \tag{8.8}$$

and η^N is the normalising constant, and

$$\begin{aligned}
\delta_1 &= 1 + \pi_1(x) + \pi_2(x) + 2\pi_3(x) + 3\pi_4(x) \\
\delta_2 &= [\pi_2(x) + 4\pi_3(x) + 9\pi_4(x)]/2 \\
\delta_3 &= [\pi_2(x) + 8\pi_3(x) + 27\pi_4(x)]/6 \\
\delta_4 &= [\pi_2(x) + 16\pi_3(x) + 54\pi_4(x)]/24
\end{aligned}$$

This distribution represents a generalised normal distribution as it contains as a special case the standard normal distribution ($\delta_3 = \delta_4 = 0$).

Defining the error terms as:

$$u = [e - \Lambda(x)]/v \tag{8.9}$$

where $\Lambda(x)$ and v are respectively the mean and the standard deviation of e, and using the transformation technique, the distribution of u is given as:

$$f(u) = exp\left[\Theta_1(x)u + \Theta_2(x)u^2/2 - u^4/4 - \eta\right], -\infty < u < \infty \tag{8.10}$$

where $\Theta_1(x)$ and $\Theta_2(x)$ are derived from:

$$\begin{aligned}
\delta_1 &= \Lambda\left(x\right)^3/v^4 - 2\Lambda\left(x\right)\Theta_2(x)/v^2 + \Theta_1(x)/v \\
\delta_2 &= -3\Lambda\left(x\right)^2/2v^4 + \Theta_2(x)/v^2 \\
\delta_3 &= \Lambda\left(x\right)/v^4 \\
\delta_4 &= -1/4v^4
\end{aligned}$$

with the remaining terms included in the normalising constant, η, given by:

$$\eta = \ln \int_{\infty}^{\infty} \exp\left[\Theta_1(x)u + \Theta_2(x)u^2/2 - u^4/4\right] du \tag{8.11}$$

The distribution given by (8.10) represents a generalised normal distribution. It can be interpreted as the error distribution from a model of the real exchange rate; see Cobb *et al.* (1983). By constraining this distribution to have zero mean, from (8.9) this implies that $\Lambda(x)$ is the mean of the distribution of the logarithm of the real exchange rate e. This generalization of the normal distribution contrasts with the standard class of linear models where it is common to assume that u is $N(0,1)$, implying that the distribution of e is $N(\Lambda(x), v^2)$.

The distributional functions $\Theta_1(x)$ and $\Theta_2(x)$ provide information which can be used in formulating the conditional expectation of e. This means that the market fundamentals can feed into the real exchange rate through two channels: a linear channel through $\Lambda(x)$, and a nonlinear channel through $\Theta_1(x)$ and $\Theta_2(x)$. It is the latter channel which provides the additional information for improving conditional forecasts of the real exchange rate and highlights the main difference between the nonlinear and linear real exchange rate models discussed above. Further specifications of $\Theta_1(x)$ and $\Theta_2(x)$ are given in Section 8.3.

8.1.3 Properties of the Model

The generalised normal distribution given by (8.10) exhibits great flexibility in modelling both skewed, kurtic and bimodal distributions. These properties give the nonlinear model potential gains in terms of explanatory and forecasting power over the linear model as well as other nonlinear models, including nonparametric and semi-nonparametric formulations.

A procedure for identifying bimodality is given by inspecting the sign of Cardan's discriminant:

$$\delta = [\Theta_1(x)/2]^2 - [\Theta_2(x)/3]^3 \tag{8.12}$$

Unimodality occurs when $\delta > 0$, while bimodality occurs when $\delta < 0$. Furthermore, a necessary condition for bimodality is that $\Theta_2(x) > 0$. The parameters of the canonical normal distribution in (8.9) have the interpretations that with unimodality $\Theta_1(x)$ is a measure of skewness and $\Theta_2(x)$ is a measure of kurtosis, while with bimodality, $\Theta_1(x)$ is a measure of the relative heights of the two modes and $\Theta_2(x)$ is a measure of the separateness of the two modes: see also Cobb *et al.* (1983).

The modes of the distribution correspond to the stable equilibria, and the antimode corresponds to the unstable equilibria, a region of relatively low probability. When there are multiple equilibria, shocks to the market fundamentals can result in discrete 'jumps' in the real exchange rate that are not necessarily reversible, giving rise to hysteresis.

The model highlights in an explicit way the role of higher-order moments. In particular, the distribution in (8.10) can model both skewness (third moment) and kurtosis (fourth moment). Where the reduced form in (8.1) is derived from an explicit structural model, these higher-order moments can be given an economic interpretation; see chapter 1. This contrasts with previous approaches in modelling empirical exchange rate distributions which have largely concentrated on statistical approaches in modelling leptokurtosis; see, for example, Friedman and Vandersteel (1982).

8.2 Estimation and Prediction

8.2.1 Maximum Likelihood Estimation

Cobb *et al.* (1983) and Lye and Martin (1993a) have shown how the parameters of distributions within the generalised exponential class can be estimated by maximum likelihood procedures. Given data on $\{e_t, x_t\}$, $t = 1, 2, ..., T$, the log likelihood is:

$$\begin{aligned} \ln L &= \sum_t \ln f(u_t) \\ &= \sum_t \left[\Theta_1(x_t)u + \Theta_2(x_t)u_t^2/2 - u_t^4/4\right] - \sum_t \eta_t \end{aligned} \tag{8.13}$$

where as in (8.9) $u_t = [e_t - \Lambda(x_t)]/v$.

Iterative procedures are required to maximise the log-likelihood function in (8.13). The Newton-Raphson algorithm in conjunction with the Berndt-Hall-Hall-Hausman algorithm are adopted here. The program MAXLIK in GAUSS is used to perform the calculations based on numerical derivatives. At each iteration, the GAUSS numerical integration program INTQUAD1 is used to compute the normalising constant and the derivatives.

8.2.2 Prediction Conventions

In most models of the exchange rate, the distribution is assumed to be unimodal. The choice of the best predictor depends upon the adopted loss function, but whatever the choice of the loss function, the set of predictors is small. If the distribution is both unimodal and symmetric, this set is reduced even further: a typical choice is the mean of the distribution.

Difficulties arise when the distribution is multimodal. Some potential new conventions are:

Global: Choose the global maximum at each point in time.

Delay: Do not leave a mode until that mode disappears.

Nearest: Choose the mode or antimode closest to the data.

The global convention is based on the assumption that the process always moves (jumps) to a region of higher probability. With the delay convention, it is assumed that a process never leaves a stable region until that region becomes unstable; see Cobb *et al.* (1983). The nearest convention consists of choosing the mode, stable or unstable, which is closest to the data. This convention allows for the possibility that the process can settle at unstable equilibria as well as stable equilibria.

8.3 Empirical Application

8.3.1 Model Specification

This section applies the above model of the distribution of the real exchange rate to the real US/UK bilateral exchange rate. The data are those used by Baillie and Pecchenino (1991). The main market fundamental variables of the real exchange rate are based on the flexible price monetarist real exchange rate model and are the domestic and foreign real money supply, domestic and foreign real income, and domestic and foreign interest rates. The assumption of flexible prices is consistent with the underlying theory of the model developed above whereby the observed real exchange rate at time t represents a realisation of a drawing from the stationary distribution at time t. The set of market fundamentals could be expanded to include other variables, while alternative definitions of the set of variables already included in the set of market fundamentals could be adopted. However, to keep the empirical analysis as comparable as possible with the Baillie and Pecchenino (1991) empirical study, these alternative variables are not considered.

The standard linear exchange rate model is embedded into the nonlinear model by expressing $\Lambda(x)$ in (8.9) as a linear function of the exogenous variables, while assuming that v is constant. More elaborate versions of this model could be considered by allowing for some heterogeneity by assuming that v is a function of either market fundamentals or follows an ARCH process, and therefore is time-varying. This generalisation of the linear model is not pursued in the empirical analysis, but will be investigated at a later stage.

A feature of the nonlinear model is that the first four moments of the distribution of the real exchange rate are time-varying. This contrasts with the standard linear model where only the first moment is time-varying, or at most the second moment is time-varying if some heterogeneity is included into the specification by making the variance time-varying. To capture changes in the characteristics of the real exchange rate distribution over time, and hence to allow for nonlinear relationships between the market fundamentals and the real exchange rate, $\Theta_1(x)$ and $\Theta_2(x)$ in (8.10) are also expressed as functions of the market fundamental variables. The model is given by:

$$\begin{aligned}
\Lambda_t &= \lambda_1 m_t^{US} + \lambda_2 m_t^{UK} + \lambda_3 y_t^{US} + \lambda_4 y_t^{UK} + \lambda_5 r_t^{US} + \lambda_6 r_t^{UK} \\
v_t &= v \\
\Theta_{1,t} &= \theta_{1,1} m_t^{US} + \theta_{1,2} m_t^{UK} + \theta_{1,3} y_t^{US} + \theta_{1,4} y_t^{UK} + \theta_{1,5} r_t^{US} + \theta_{1,6} r_t^{UK} \\
\Theta_{1,t} &= \theta_{2,1} m_t^{US} + \theta_{2,2} m_t^{UK} + \theta_{2,3} y_t^{US} + \theta_{2,4} y_t^{UK} + \theta_{2,5} r_t^{US} + \theta_{2,6} r_t^{UK} \\
u_t &= \left(e_t - \Lambda_t\right)/v \qquad (8.14)
\end{aligned}$$

where m is the logarithm of the real money stock, y is the logarithm of real income, r is the nominal interest rate, and e is the logarithm of the exchange rate.

The sample period consists of 207 monthly observations starting in March 1973, when the US floated its exchange rate, and ending in May 1990. The real exchange rate is computed from the nominal exchange rate, which is the end-of-month closing bid price on the New York foreign exchange market, and the CPI in the two countries. The stock of money is M1 for the US and M0 for the UK. The real income variables are the industrial production indexes of the US and the UK. The US interest rate is the Federal Funds rate, while for the UK it is London Interbank Offer. All data are adjusted for seasonality by regressing each series on a constant and three seasonal dummy variables and using the residuals as the seasonally adjusted data. Finally, to avoid numerical overflows when computing the integrals in (8.11) numerically, all data are standardised to have unit variance.

8.3.2 The Results

Maximum likelihood parameter estimates and diagnostics of the nonlinear model are given in Table 8.1. For comparison, ordinary least squares estimates of the linear real exchange rate model are also given. The nonlinear model performs very well when compared with the linear model. In the case of the nonlinear model, the signs of the parameter estimates of the function which determines the mean of the distribution, $\Lambda(x)$, are consistent with the basic monetarist theory of real exchange rates. This contrasts with the linear model where half of the parameter estimates have the wrong sign. There is a big improvement in goodness of fit with $\overline{R}^2$ increasing from 0.282 for the linear model to 0.736 for the nonlinear model. Both the AIC and SIC show that the nonlinear model is superior to the linear model, while the Keenan test shows that the nonlinear model captures all of the nonlinearities in the real exchange rate. For a discussion of the properties of the Keenan test, see Lee *et al.* (1993).

The improvement in explanatory power of the nonlinear model over the linear model is highlighted in Figure 8.1 which compares the predictions of the two models with the actual logarithm of the real exchange rate. There are two periods worth highlighting: 1980-81 when the real exchange rate peaked, and 1985 when it hit a trough. In both cases, the linear model fails to capture the large swings in the real exchange rate.

8.3.3 Forecasting

Meese and Rogoff (1983) found that the simple random walk specification performed better than many linear models of exchange rates. This work was extended by Meese and Rose (1991) who found that the random walk model also performed better than a number of nonlinear models. In contrast Engel and Hamilton (1990) found that a nonlinear model based on stochastic segmented trends yielded superior forecasting power to the random walk. To test the forecasting properties of the nonlinear exchange rate model developed here, its forecasting capabilities are compared with these existing methodologies.

One way to highlight the relationship between the nonlinear model developed here with that of the random walk model is to express the error correction model given by (8.3) in terms of the logarithm of the exchange

Table 8.1: Estimates of the real exchange rate model, 1973:3 to 1990:5

Variable	Parameter	Linear		Nonlinear	
		Coef.	(t-stat.)	Coef.	(t-stat.)
m^{US}	λ_1	-0.066	(0.485)	0.081	(0.503)
m^{UK}	λ_2	0.178	(1.349)	-1.489	(7.628)
y^{US}	λ_3	-0.150	(0.563)	-1.679	(7.005)
y^{UK}	λ_4	0.470	(2.293)	0.216	(1.202)
r^{US}	λ_5	0.411	(3.991)	0.525	(4.919)
r^{UK}	λ_6	0.168	(2.282)	-0.245	(3.011)
m^{US}	$\theta_{1,1}$			-0.595	(0.694)
m^{UK}	$\theta_{1,2}$			4.683	(4.208)
y^{US}	$\theta_{1,3}$			5.145	(3.683)
y^{UK}	$\theta_{1,4}$			-0.021	(0.020)
r^{US}	$\theta_{1,5}$			-0.005	(0.014)
r^{UK}	$\theta_{1,6}$			0.451	(1.190)
m^{US}	$\theta_{2,1}$			-4.349	(7.279)
m^{UK}	$\theta_{2,2}$			0.959	(2.052)
y^{YS}	$\theta_{3,1}$			5.502	(5.477)
y^{UK}	$\theta_{4,1}$			-0.176	(0.281)
r^{US}	$\theta_{5,1}$			-1.925	(5.309)
r^{UK}	$\theta_{6,1}$			0.128	(0.599)
	v	0.847		1.034	(22.299)
Diagnostics					
$\overline{R}^2$		0.282		0.736	
RSS		144.280		49.807	
$\ln L$		-256.360		-132.265	
SIC		544.717		360.520	
K		6.915		0.430	

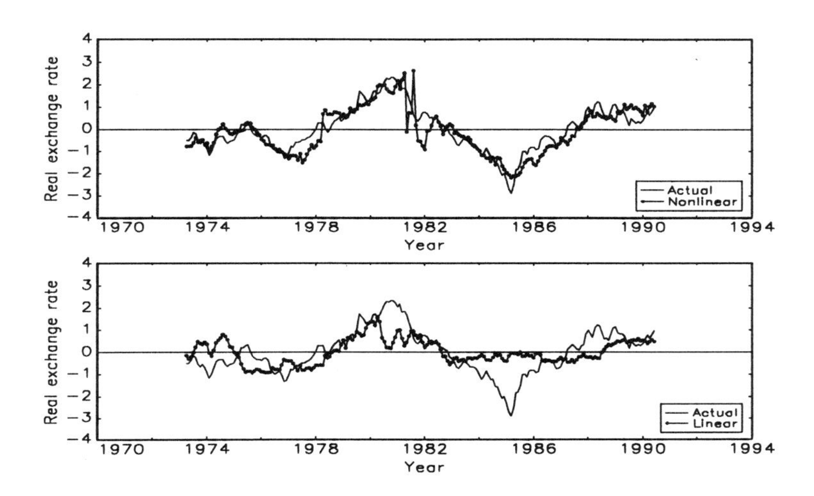

Figure 8.1: Comparison of linear and nonlinear predictions

rate, $e = log(s)$, and assume that $\mu(e, x) = \alpha_0$, and $\sigma^2(e, x) = \beta_0$, are constant:

$$de = -\alpha_0 dt + \beta_0 dW \tag{8.15}$$

It is possible to derive a closed-form expression for the transitional density defined by (8.4). Using the results of Cox and Miller (1984, p.209), the transitional density is

$$f(e_t) = \frac{1}{(2\pi\beta_0 t)} \exp\left[-\frac{(e_t - \alpha_0 t)^2}{2\beta_0 t}\right] \tag{8.16}$$

This is simply a normal distribution of the level of the (logarithm of) the exchange rate with mean $\alpha_0 t$, and variance $\beta_0 t$. These moments are also the moments of a random walk with drift.

Equation (8.5) suggests that one way to compare the random walk specification with the nonlinear model is to compute the RMSE obtained from the predictions of a model based on regressing the exchange rate on a constant and a time trend. This result is represented in Table 8.2 by RW(level)=0.995, which is larger than the one-step-ahead RMSE of the nonlinear model based on the level of the exchange rate given by NL(level)=0.491.

An alternative linear model of the level of the exchange rate is based on the (logarithm of the) forward exchange rate at time $t-1$ for delivery at time

Table 8.2: RMSE based on predictions

Model	RMSE
RW(level) random walk based on (8.16)	0.995
NL(level) nonlinear model based on (8.14)	0.491
SE(level) speculative efficiency, (8.17) $\alpha = 0$ and $\beta = 1$.	0.254
RW(sr) random walk model using (8.18) with $\phi_i = 0, \forall i$	0.193
NL_1(sr) the ECM in (8.18) with $\phi_4 = 0$	0.190
NL_2(sr) The ECM in (8.18) unrestricted	0.183
EH(sr) Engel and Hamilton model	0.194

t, f_{t-1} . Under the assumption of speculative efficiency, $\alpha = 0$ and $\beta = 1$ in the following regression equation (see Baillie and McMahon, 1990, for a review):

$$s_t = \alpha + \beta f_{t-1} + \xi_t \tag{8.17}$$

where ξ_t is a disturbance term. This suggests that a one-step-ahead prediction of the exchange rate is given by f_{t-1} . Using one-month forward exchange rate data between the US and the UK, the RMSE of this forward rate, assuming speculative efficiency, is given by SE(level) = 0.254 in Table 8.2. As this RMSE is smaller than that obtained using the nonlinear model given by NL(level) in Table 8.2, this suggests that the information set used in formulating forecasts of the spot rate in forward markets is larger than the set of variables used to estimate the nonlinear long-run model.

The forecasts of the nonlinear model presented in Table 8.2, discussed so far, are based on the stationary distribution and can thus be interpreted as long-run forecasts. However, it is more appropriate to generate the forecasts from the error correction representation as given by equation (8.3) and hence focus attention on the change in the exchange rate, that is, short-run forecasts. This has the further advantage that the forecasts of the nonlinear model become commensurate with forecasts obtained using the nonlinear stochastic segmented trends model which is also based on the change in the exchange rate. Having derived forecasts of the change in the exchange rate, forecasts of the level of the exchange rates can be obtained by progressively summing the forecasts of the changes.

In building an error correction model for the nonlinear model several alternative specifications are possible. The approach adopted here is to estimate the following ECM:

$$e_t - e_{t-1} = \phi_1 \Theta_{1,t-1} + \phi_2 \Theta_{2,t-1} u_{t-1} + \phi_3 u_{t-1}^3 + \phi_4 \left(\Lambda_t - \Lambda_{t-1} \right) + \omega_t \tag{8.18}$$

where ω_t is an error term and $\Theta_{1,t}$, $\Theta_{2,t}$, Λ_t and u_t are defined in (8.14). The motivation behind (8.18) is that it represents a discretisation of the continuous time ECM expression used to derive the stationary distribution for e_t. The first three variables in (8.18) correspond to the terms used to identify the modes of the distribution in (8.10) and thereby correspond to μ, in an error correction model of e in the form given by (8.3). The fourth term, $\Lambda_t - \Lambda_{t-1}$, is used to capture any short-term dynamics which have not been captured by the error-correction term. While further short-run dynamics could be envisaged, the variable $\Lambda_t - \Lambda_{t-1}$ is sufficient for the purposes of this chapter to highlight the forecasting properties of the nonlinear model. Another approach is to estimate the continuous time version of the ECM by using the indirect estimation method of Gourieroux *et al.* (1993). This was not tried, but will be investigated in future work.

The short-run RMSE of estimating the nonlinear ECM model in (8.18) with $\phi_4 = 0$ is given by $NL_1(sr) = 0.190$ in Table 8.2. This is marginally smaller than that obtained for the random walk representation, $RW(sr) = 0.193$, which is simply the standard deviation of $e_t - e_{t-1}$. Not constraining ϕ_4 to be zero in (8.18) yields a RMSE of $NL_2(sr) = 0.183$, which constitutes just over a 5 per cent reduction in the RMSE compared with that obtained for the random walk representation.

As a final comparison, the forecasting performance of a two-state stochastic segmented trends model adopted by Engel and Hamilton (1990) is examined. This comparison is of special interest for the present chapter because the stochastic segmented trends model, as with the nonlinear model developed here, is based on the assumption that the distribution governing the exchange rate is also bimodal. The one-step-ahead predictions are based on Engel and Hamilton (1990, p.699). The RMSE is given by $EH(sr) = 0.194$, in Table 8.2, which is marginally larger than the RMSE using the random walk, which is in turn larger than that obtained using the nonlinear model.

8.3.4 Interpretation

The large movements in the real exchange rate during the periods 1980 and 1985 can be interpreted as resulting from multiple equilibria; see Lye and Martin (1994) who found similar distributional characteristics for the rate of growth of the nominal $US/$A exchange rate. This is seen from Figure 8.2, which shows that Cardan's discriminant, as given by equation (8.13), becomes negative for these periods. The changing distributional properties of the real exchange rate over the sample period are further highlighted in Figures 8.3. It is interesting to note that the within-sample predictions of the linear model in 1985, as given in Figure 8.1, are closer to the antimode of the distribution. As this corresponds to a region of relatively low probability,

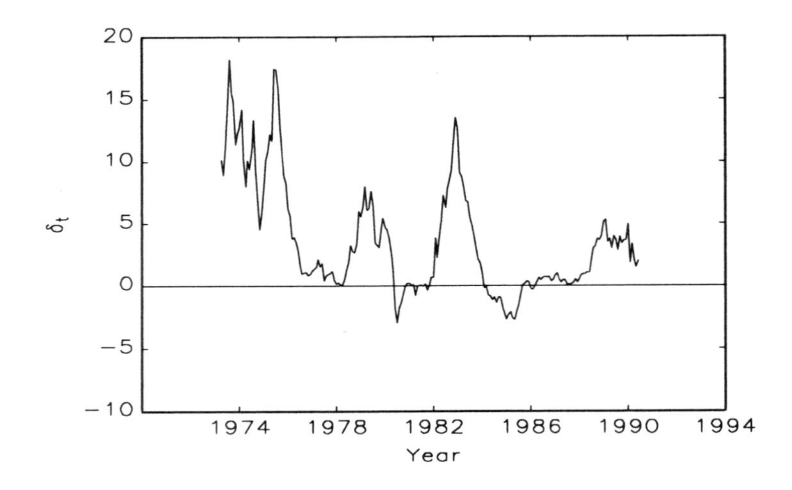

Figure 8.2: Measure of bimodality: $\delta_t < 0$ implies bimodality

it helps to highlight further the problems of explaining real exchange rate movements with a linear model.

An alternative explanation of the occurrence of discrete jumps around the years 1980 and 1985 is that they were the result of structural changes. For example, there was a change in the Federal Reserve Board operating system around 1980. This property highlights further the flexibility of the modelling framework adopted in this chapter. The usual approach to modelling structural change is by augmenting the model with dummy variables. This is a mechanical approach as the timing of the change is determined exogenously. In contrast, the real exchange rate model developed above enables the the timing of the structural change to be endogenised.

The occurrence of bimodality has important implications for conducting policy and highlights some of the potential problems that monetary authorities may have in correctly choosing their policy settings. For example, given that the change in the way the Federal Reserve Board conducted its monetary policy in 1980 occurred during a period when the real exchange rate was in a period of bimodality, small changes in monetary settings could have been enough to generate large changes in the real exchange rate. This contrasts with other periods when the real exchange rate was operating in a unimodal zone, whereby the effect on the real exchange rate from the same change in

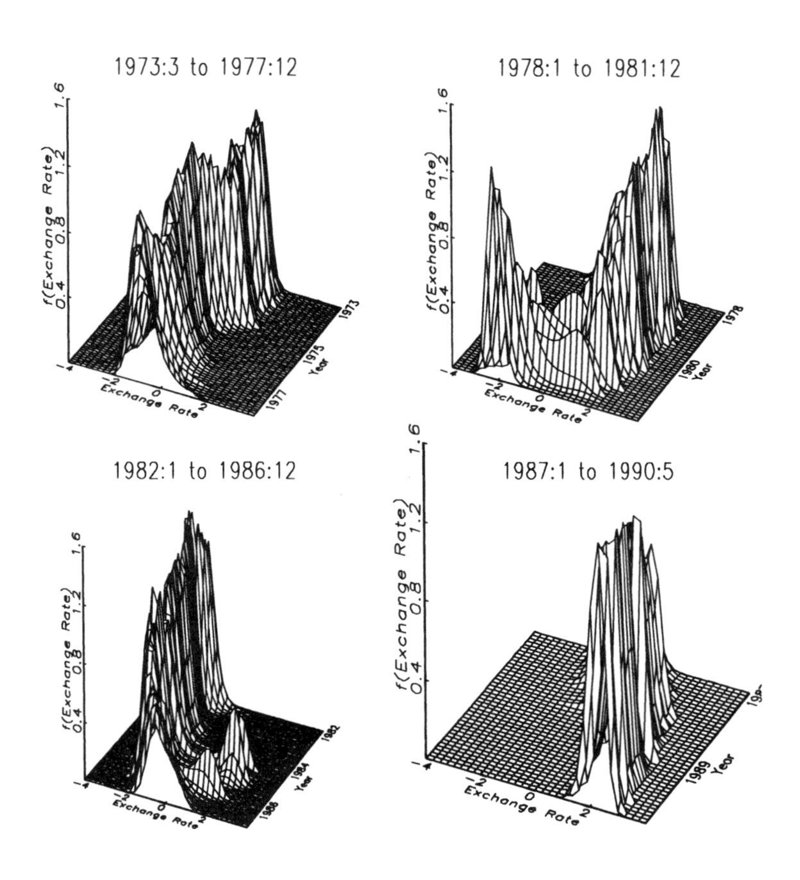

Figure 8.3: US/UK exchange rate distributions: 1973:3 to 1990:5

monetary policy would be expected to be less.

8.4 Conclusions

This chapter has provided a framework for nonlinear models of the real exchange rate. The approach is based on deriving the stationary density as the solution of a continuous-time error correction model. An important feature of this density is that it nests the standard normal distribution, as well as generating a range of distributional characteristics such as skewness, kurtosis, and bimodality.

The framework was used to estimate a nonlinear model of the real US/UK exchange rate for the period from 1973 to 1990 using monthly data. The nonlinear model was compared with a linear model and found to be superior in terms of goodness of fit and explaining the nonlinearities in the data. The nonlinear model was also found to be superior to the random walk representation and the stochastic segmented trends model of Engel and Hamilton, yielding over a 5 per cent reduction in RMSE. The reason for this superiority was that the nonlinear model was able to identify periods when the distribution became bimodal. These periods corresponded to large swings in the real exchange rate which were caused by structural changes not explained by the linear model. The identification of structural changes in the real exchange rate provides an explanation of the slow adjustment of the exchange rate to purchasing power parity; see, for example, the approach adopted by Pesaran and Shin (1994). The nonlinear model was also able to explain why the linear model performed badly: during periods of bimodality it generated predictors corresponding to the antimode of the distribution.

Two important areas of further research stem from this paper. First, there are some obvious connections between the model developed above and the cointegration framework of Engle and Granger (1987). If the stationary distribution is interpreted as a long-run distribution then this represents the distributional form of cointegration; see Creedy *et al.* (1994). Furthermore, given the emphasis on solving nonlinear models, this implies that the model derived here constitutes an appropriate framework for estimating and testing nonlinear cointegrating models. This connection with cointegration gives rise to another important point: a failure to reject the null hypothesis of no cointegration does not necessarily mean a rejection of the long-run economic model, but a rejection of the functional form. For example, in the empirical work of Baillie and Pecchenino (1991) where the assumption of purchasing power parity was rejected, this could be interpreted as a rejection of the linear model used in testing for cointegration rather than a rejection of the economic fundamentals driving the long-run model. Furthermore, the model developed here could be viewed as a generalisation of the work of Baillie and Pecchenino as it extends the linear cointegration framework of these authors to nonlinear cointegration.

A second area of research suggested is that it would be useful to compare the properties of the exchange rate model developed here with other flexible procedures based on nonparametric and semi-parametric methods discussed by Meese and Rose (1991); also see Chinn (1991). Given the limited success that Meese and Rose found with the nonparametric and semi-nonparametric models, it is hypothesised that the nonlinear approach adopted in this chapter can give rise to potential gains from an explanatory and forecasting perspective. These potential gains arise from explicit formulation of higher-order moments of the distribution which contrasts with these other approaches where either higher-order moments are treated implicitly or attention is focused only on the mean of the distribution. Potential gains also arise from the ability of the model to exhibit multimodality, which contrasts with nonparametric approaches where the distribution is unimodal. This property is particularly useful in explaining large swings in the exchange rate as well as persistent deviations.

Chapter 9

Jump Models and Higher Moments

G.C. Lim, Jenny N. Lye, Gael M. Martin and Vance L. Martin

The use of dummy variables in regression analysis constitutes an important statistical procedure for modelling structural change; see Hackl (1989). Although this approach may be justified for some empirical applications, in general it represents an *ad hoc* mechanism that potentially masks the underlying dynamics of the process being modelled. Questions concerning why processes exhibit jumping behaviour need to be addressed. In the past modellers have tended to dodge the more difficult episodes in time when economic variables exhibit large fluctuations, by introducing dummy variables into the linear models. Whilst these points in time are difficult to model, they are clearly more interesting and represent a challenge.

Models of price bubbles provide an example. In the work of Flood and Garber (1980), the period of the price bubble is modelled by using a dummy variable with a time trend, where the timing of the bubble is determined from inspection of the data. This model does not endogenise the determination of the timing of the bubble, nor does it address formally why the bubble bursts. If bursting bubbles are the result of prices operating in a zone of multiple equilibria whereby some equilibria are dominated by self-fulfilling expectations whilst other equilibria are dominated by market fundamentals, the use of dummy variables in a regression equation that relates price to a set of market fundamental variables in a linear way misses the point.

The aim of this chapter is to investigate a class of models which caters for jumping behaviour in a more natural way. The approach is based on the generalised exponential family of distributions, introduced in chapter 1. This class of distributions can exhibit multimodality when the underlying processes exhibit multiple equilibria. An advantage of this framework is that jumping behaviour can be modelled without the aid of dummy variables, and the timing of jumps is determined endogenously rather than imposed exogenously.

A general class of models for capturing jump dynamics is developed in Section 9.1. The key property of this class of models is that the inclusion of higher order moments into the model not only allows for distributional features such as skewness and kurtosis, it also allows for multimodality. To highlight the properties of this framework, simple step functions and heteroscedastic processes are simulated from higher-order moment models in Section 9.2. An empirical example of jumping is given in Section 9.3 where the higher order moment model is applied to modelling the Australian stock market crash in 1987.

9.1 A General Model of Jump Dynamics

This section presents a general model of jump dynamics. The approach consists of constructing switching processes within classes of distributions which can exhibit multimodality. A useful starting point is to consider the following dynamic model of the continuous time process Y (see Cox and Miller, 1984):

$$dY = -\mu\left(Y, X\right)dt + \sigma\left(Y, Z\right)dW \tag{9.1}$$

where dW is a Wiener process which is distributed as $N(0, dt)$, and X and Z represent respectively sets of explanatory variables which affect the mean of dY, that is $\mu(.)$, and the variance of dY, that is $\sigma(.)$. Equation (9.1) is simply an Ito process which can be interpreted as a continuous time error correction model (see Phillips, 1991c). That is, a positive shock, $dW > 0$, results in an increase in Y in the short-run, but over time Y error corrects back to its long-run value. The countervailing force is identified by the negative sign in front of the mean $\mu(.)$ in (9.1). Since Y, and by implication X, are integrated processes, $\mu(.)$ can also be interpreted as a cointegrating error in the sense of Engle and Granger (1987).

The equilibrium solution of the density of Y is given, as in chapter 1, by:

$$f\left(y\right) = \exp\left[-\int_0^y \left(\frac{2\mu(w) + d\sigma^2\left(w\right)/dw}{\sigma^2\left(w\right)}\right)dw - \eta\right] \tag{9.2}$$

where η is the normalising constant such that $\int f(y)dy = 1$. This density constitutes the stochastic analogue of cointegration. However, it is a general expression which depends upon explicit representations of $\mu(.)$ and $\sigma^2(.)$ given in (9.1).

9.1.1 The Linear Model

To help motivate the general framework for modelling jumping behaviour pursued in this chapter, the linear model, together with some of its modifications, is derived initially. For the linear model, the mean is specified as:

$$\mu(Y, X) = Y - \alpha X \tag{9.3}$$

where α is a set of parameters. In the case where the linear model has constant variance, $\sigma^2(.)$ is given by:

$$\sigma^2(.) = \beta \tag{9.4}$$

If the linear model exhibits heteroscedasticity, (9.4) is generalised to:

$$\sigma^2(.) = \beta(Z) \tag{9.5}$$

Using the expressions for the mean and the variance in (9.3) and (9.4) respectively in equation (9.2) gives the standard normal distribution:

$$f(y) = \exp\left[\sum_{i=1}^{2} \Theta_i y^i - \eta^N\right], -\infty < y < \infty \tag{9.6}$$

where η^N is the normalising constant. In the usual formulation of the normal distribution:

$$\begin{aligned} \Theta_1 &= \mu_y / \sigma_y^2 \\ \Theta_2 &= -1/2\sigma_y^2 \end{aligned}$$

where μ_y and σ_y^2 are the mean and the variance of Y respectively.

9.1.2 The Nonlinear Model

Suppose now that the relationship between Y and X is given by the general implicit functions $\mu(.)$ and $\sigma^2(.)$. One way to derive an explicit form of the density of Y is to expand these functions in a Taylor series expansion around some appropriately chosen point. For $\mu(.)$ the expression becomes:

$$\mu(Y, X) = \sum_{i=0}^{M-1} \alpha_i(X) y^i \tag{9.7}$$

The choice of M in (9.7) is based on the underlying theory. For example, setting $M = 4$ in (9.7) is appropriate in the general equilibrium model of international trade investigated by Creedy and Martin (1993). Setting $M =$

4 is also appropriate for capturing the qualitative properties of the intra-industry trade model of Venables (1984). A common feature of both of these models is that they can generate multiple equilibria. Finally, the linear model is clearly a special case of (9.7) and can thus be interpreted as a local, first order approximation of a nonlinear model around some (equilibrium) point.

The density obtained when (9.7) and (9.4) are substituted in (9.2) yields the generalised normal distribution:

$$f(y) = \exp\left[\sum_{i=0}^{M} \Theta_i y^i - \eta\right], \quad -\infty < y < \infty \tag{9.8}$$

where η is the normalising constant. The most striking feature of the density in (9.8) when compared with the standard normal distribution is the higher order terms, that is, terms of order greater than two. These higher order terms result in this density having higher order moments that are independent of the first two moments. This contrasts with the normal distribution where all higher order moments are a simple function of the first two moments.

An important property of this density is that it can exhibit a wide range of distributional shapes. For example, it can display skewness, a property that is characteristic of the distribution of growth cycles (see Neftci, 1984), and kurtosis, which is a feature of exchange rate distributions (see, for example, Friedman and Vandersteel, 1982). More importantly, the generalised normal distribution can display multimodality, with the number of modes determined by the number of roots of the polynomial in (9.7). For example, if $M = 4$ in (9.7), there are at most three roots and (9.8) is bimodal. Two of these roots are stable and correspond to the modes of (9.8), whilst the unstable root corresponds to the antimode of (9.8). There is a one-to-one relationship between multiple equilibria and higher order moments, and for this reason multiple equilibria cannot be modelled by using a unimodal distribution.

The interesting dynamical properties of this model arise when, through changes in X, the process moves from a state of a unique equilibrium with a corresponding unimodal density, to a state of multiple equlibria where the density exhibits multimodality, then back to a state of unique equilibrium. It is in these situations that jumping occurs, with the movement from one stable equilibrium to another. This jumping behaviour is characterised in the next section.

9.2 Theoretical Examples

To highlight some of the properties of the generalised normal distribution given by (9.8), a number of Monte Carlo experiments are undertaken. In each of the simulation experiments, random numbers are drawn from a generalised normal distribution by the inverse cumulative distribution technique. The programs are written using GAUSS.

9.2.1 Step Function

This example shows how a discrete change in a variable can be the result of jumping from one stable equilibrium to another. The approach is based on the generalised normal distribution:

$$f(y) = \exp\left[\Theta_1 y + 1.0y^2 - 0.05y^4 - \eta\right] \tag{9.9}$$

where η is the normalising constant and Θ_1 is allowed to vary from -5 to 5 in steps of 0.2. For each of the 51 values of Θ_1, a single realisation of Y is drawn from the density in (9.9). For large negative values of Θ_1, the density is unimodal. As Θ_1 approaches zero, another mode appears which becomes the dominant mode when Θ_1 becomes positive.

The results of this experiment are given in Figure 9.1 and show that Y is trending upwards with a structural break around $\Theta_1 = 0$. Clearly the usual approach of inspecting a time series plot of the data and plugging the gap with a shift dummy variable is inappropriate for the present example since this strategy hides the true dynamics of the process; namely the structural break is the result of multiple equilibria and the process switching from one stable equilibrium to another.

To highlight the problems of using dummy variables to model structural breaks, the results of fitting a linear model to the data in Figure 9.1 are as follows:

$$\begin{aligned}\widehat{Y} &= \underset{(22.546)}{-0.267} + \underset{(4.036)}{0.043}\, X + \underset{(18.837)}{5.962}\, D \\ \overline{R}^2 &= 0.976\end{aligned}$$

where absolute values of t-statistics are given in brackets, X is a 'time' trend, and D is a dummy variable which equals zero for the first 25 observations and unity thereafter. These results show that this model does a very good job at masking the underlying dynamics as it yields too good a fit!

9.2.2 Heteroscedasticity

Heteroscedasticity arises when the second moment of a distribution is not constant over the sample period, but is a function of a set of variables. The simplest way to identify heteroscedasticity is to plot the series over the sample period and see if the series begins to gyrate. An alternative explanation

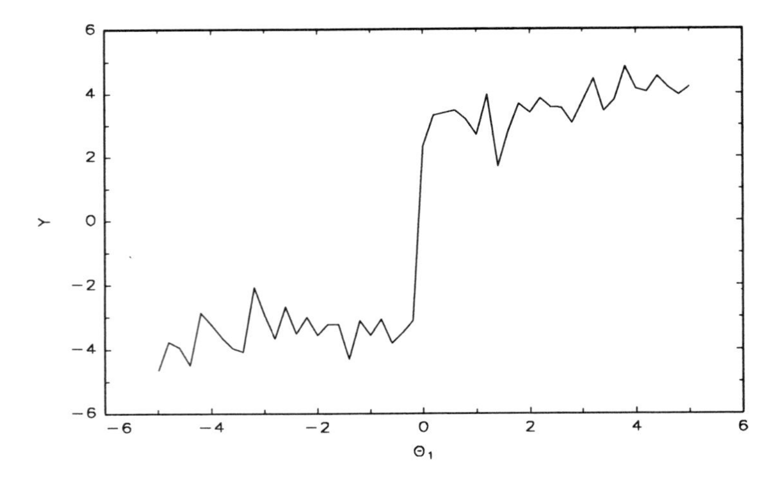

Figure 9.1: Simulation of a jump process.

of this phenomenon is that the underlying density is bimodal, or even multimodal, and that the modes are moving further apart over the sample period. To highlight this point, consider the following generalised normal distribution:

$$f(y) = \exp\left[\Theta_2 y^2 + 0.25y^4 - \eta\right] \tag{9.10}$$

where η is the normalising constant and Θ_2 is allowed to vary from -5 to 5 in steps of 0.2. The qualitative properties of the density in (9.10) are that there is a single mode for negative values of Θ_2, whereas there are two modes for positive values of Θ_2 which are pushed further away the larger Θ_2 is. This model is known as a bifurcation model where the bifurcation point occurs at $\Theta_2 = 0$; see Cobb *et al.* (1983).

The experiment in this example proceeds as before: for each of the 51 values of Θ_2, a single realisation of Y is drawn from the density in (9.10). The results of the experiment are given in Figure 9.2 and show evidence of increasing variability in the variable Y. However, it would be incorrect to interpret the increasing variability in Y as the result of heteroscedasticity since both second and fourth order moments are important and not just the second moment.

To highlight the problems of using the standard tools, a regression equation is estimated by regressing Y^2 on a 'time' trend. The estimated regression

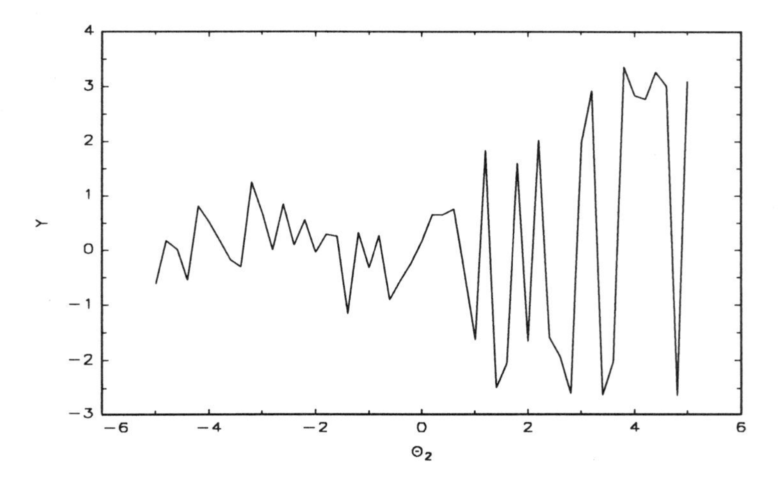

Figure 9.2: Simulation of heteroscedasticity

line is:

$$\widehat{Y}^2 = \underset{(4.048)}{-2.198} + \underset{(10.256)}{0.186}\ X \tag{9.11}$$
$$\overline{R}^2 = 0.676$$

where absolute values of t-statistics are given in brackets, and X is a 'time' trend. As the results show, there is strong evidence of a time-varying second moment. However, these results do not show or provide any information on the reasons for the nonstationarity.

In addition, if a test for first order ARCH is performed on the variable Y, the calculated value is 36.207, which is highly significant when compared with the critical value $\chi_1^2\,(0.05) = 3.84$. This suggests that the draws of Y are dependent on lagged values of Y when in fact they are not. This result also highlights the problems of using ARCH models to explain volatility, as these models are based on time-varying second moments which provides little insight into dynamics arising from time-varying higher order moments.

9.3 The Australian 1987 Crash

In this section, the theoretical framework discussed above is applied to modelling the Australian stock market crash in October 1987. This example is motivated by seeing if a model that explicitly takes into account higher order moments can model the large fall in the stock price that occurred on 20 October, 1987.

The data are daily starting from 2 January 1986 and ending on 30 December 1988. The total number of observations is 759. The share price data are the all ordinaries index, P_t. The dividend data d_t, are computed from the accumulation index. All data are obtained from the Stock Market database.

9.3.1 Integration and Cointegration Properties

The data, P_t and d_t, are initially tested for unit roots by using the structural break unit root test of Perron and Vogelsang (1992). This test is an extension of the augmented Dickey-Fuller (ADF) (see Fuller, 1976, and Phillips, 1987) unit root test to the case where the underlying process can exhibit a structural break in the intercept at an unknown point in time. The calculated values of the test statistic are -2.003 and -1.878 for P_t and d_t respectively. Given a 5 per cent critical value -4.44 (see Perron and Vogelsang, Table 1, $T = \infty$, 1992), the null hypothesis of a unit root is not rejected at the 5 per cent level.

Given the unit root properties of P_t and d_t, the Gregory and Hansen (1996) cointegration test in the presence of a structural break is performed. The ADF version of the test is used which allows for a structural break in both the intercept and slope parameters. The calculated value of the test statistic is -8.234. Given a 5 per cent critical value of -4.95 (see Gregory and Hansen, Table 1, 1996), the null hypothesis of no cointegration between P_t and d_t is rejected at the 5 per cent level. The implied estimate of the timing of the break is 20 October 1987, the day that the crash occurred.

For comparison, Bayesian techniques are used to produce posterior inferences regarding both the existence of cointegration and the timing of the break. Following Martin (1996), the triangular cointegration model of Phillips and Loretan (1991) is specified for P_t and d_t, with both an intercept and slope shift accommodated. The cointegrating error is parameterized as an $AR(1)$ process with coefficient ρ. Based on the marginal posterior density of ρ, the probability of ρ being less than unity measures the probability of cointegration. The marginal posterior mass function of the discrete change point parameter r, is used to produce a modal estimate of the timing of the shift in the cointegrating relationship.

In computing the posterior densities for ρ, a marginal Jeffreys prior is

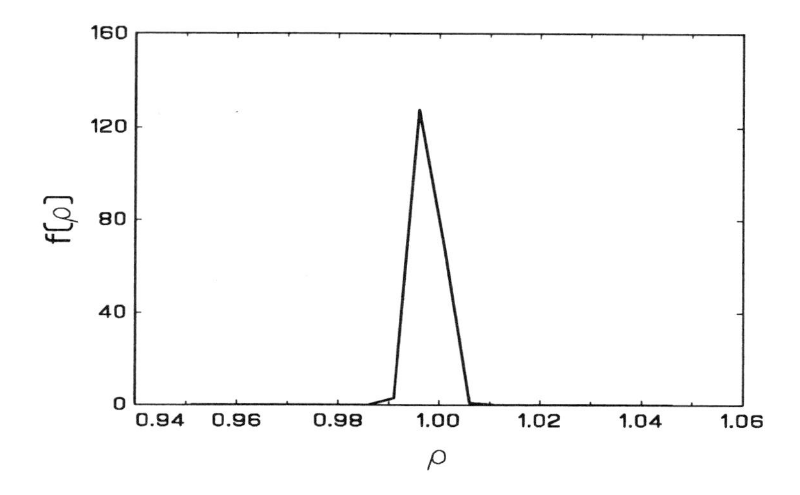

Figure 9.3: Marginal posterior of ρ

specifed for ρ, with a view to avoiding the implicit bias towards cointegration associated with a uniform marginal prior; see Phillips (1991a, 1991b and 1993). A conditional Jeffreys prior for the regression parameters, given ρ and r, serves to counteract the distortion to inferences which results from a near identification problem in the region in which ρ approaches one; see Martin (1996) for details, as well as Kleibergen and van Dijk (1994 and 1996) and Chao and Phillips (1996) for related work. The marginal prior specified for r is uniform.

The marginal posterior of ρ is reproduced in Figure 9.3. The marginal posterior is produced using a combination of analytical and low dimensional numerical integration. Despite a modal estimate of 0.996 for ρ, the degree of concentration of the marginal ρ posterior is such that the probability of cointegration is still reasonably high, at 65.3 per cent. The mass function for the timing of the break point r is estimated over a grid of 120 daily time periods around the time of the crash. This function, although not presented, ascribes 100 per cent probability to a break in the relationship between P_t and d_t on October 20, 1987.

9.3.2 Model Specification

The results of the cointegration analysis of the two series P_t and d_t show that a relationship between P_t and d_t exists provided that a set of dummy variables are included in the model to capture the timing of the crash. As an alternative modelling strategy for capturing the stock market crash, informaton on the higher order moments of P_t are used to capture the jump in P_t. An advantage of this modelling approach, as compared with the dummy variable models, is that it is possible to formulate forecasts of both the size and the timing of a crash.

A prototype empirical model is specified as:

$$
\begin{aligned}
P_t &= \lambda_1 d_t + \lambda_2 \Delta P_{t-1} + \lambda_3 \Delta d_{t-1} + \upsilon u_t && (9.12)\\
f(u_t) &= \exp\left[\Theta_{1,t} u_t + \Theta_{2,t} u_t^2/2 - u_t^4/4 - \eta_t\right] && (9.13)\\
\Theta_{1,t} &= \pi_{11} d_t + \pi_{12} \Delta P_{t-1} + \pi_{13} \Delta d_{t-1} && (9.14)\\
\Theta_{2,t} &= \pi_{21} d_t + \pi_{22} \Delta P_{t-1} + \pi_{23} \Delta d_{t-1} && (9.15)
\end{aligned}
$$

Equation (9.12) shows that the mean of the stock price is a function of the market fundamental variable d_t, as represented by λ_1. This represents the direct link between prices and dividends. The change variables ΔP_{t-1}, Δd_{t-1} are included in (9.12) to take into account the short-run market fundamental dynamics. This is also partly motivated by current work on estimating cointegrating equations whereby the inclusion of such variables can result in asymptotically efficient parameter estimates; see Phillips and Loretan (1991) and Saikkonen (1991), as well as Lim and Martin (1995) for a survey. The error term is given by u_t, with corresponding standard deviation υ, which for simplicity is assumed to be a constant. The distribution of u_t is assumed to have a generalised normal form given by (9.13). The shape of this distribution over time is determined by $\Theta_{1,t}$ and $\Theta_{2,t}$ in equations (9.14) and (9.15) respectively. For simplicity, the determinants of these functions are assumed to be the same as the mean. These terms provide the indirect linkages between the stock price and dividends.

9.3.3 Distributional Characteristics

Maximum likelihood parameter estimates of the model are given in Table 9.1 for the full sample period. The data are scaled to have zero mean and unit variance. For comparison, ordinary least squares estimates of the linear model are also given.

The distributional properties of the nonlinear model are highlighted in Figure 9.4, which gives point estimates of Cardan's discriminant over the

Table 9.1: Linear and nonlinear model parameter estimates: MLE[(a)]

Variable	Parameter	Linear	Nonlinear
d_t	λ_1	-0.549	-0.562
		(17.971)	(32.962)
ΔP_{t-1}	λ_2	0.902	-0.371
		(2.520)	(1.616)
Δd_{t-1}	λ_3	-0.109	0.606
		(0.337)	(2.453)
d_t	π_{11}		0.009
			(0.249)
ΔP_{t-1}	π_{12}		0.913
			(1.628)
Δd_{t-1}	π_{13}		-0.997
			(1.600)
d_t	π_{21}		-1.985
			(21.593)
ΔP_{t-1}	π_{22}		-0.092
			(0.081)
Δd_{t-1}	π_{23}		2.067
			(2.159)
	υ		0.833
			(45.884)

(a) t-statistics are in brackets.

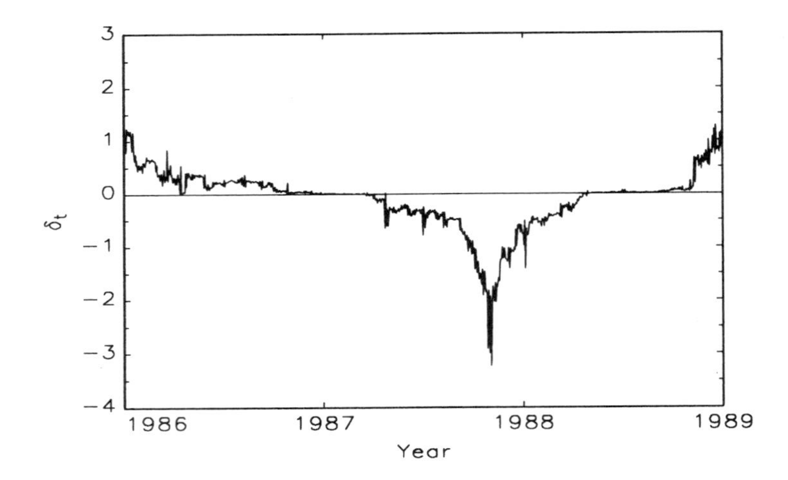

Figure 9.4: Cardan's discriminant as a test of bimodality

sample period:

$$\delta_t = \left(\frac{\Theta_{1,t}}{2}\right)^2 - \left(\frac{\Theta_{2,t}}{3}\right)^3 \tag{9.16}$$

A value of $\delta_t > 0$ ($\delta_t < 0$) signifies that the distribution is unimodal (bimodal). A necessary condition for bimodality is $\Theta_{2,t} > 0$. Inspection of Figure 9.4 reveals that the sample period can be divided into three periods. The first period is from 1986 to the first part of 1987, where $\delta_t > 0$. This is a period of unimodality and is interpreted as a time when the market has a single view of the price of the share market. The second period is when $\delta_t < 0$, which lasts for the rest of 1987 and ends in the first part of 1988. This is a period of bimodality which is interpreted as a period when the market has a dual view of the share price. In terms of the bubbles literature, these two views are the market fundamental view and the self-fulfilling expectations view; see, for example, Flood and Garber (1980). The third period occurs until the end of the sample period in 1988, when the distribution returns to being unimodal.

The predictive properties of the nonlinear model are shown in Figure 9.5 and are compared with the predictions of the linear model obtained by estimating (9.12) by OLS. The most striking result is that the nonlinear model captures the timing of the large fall in the stock price. It is worth

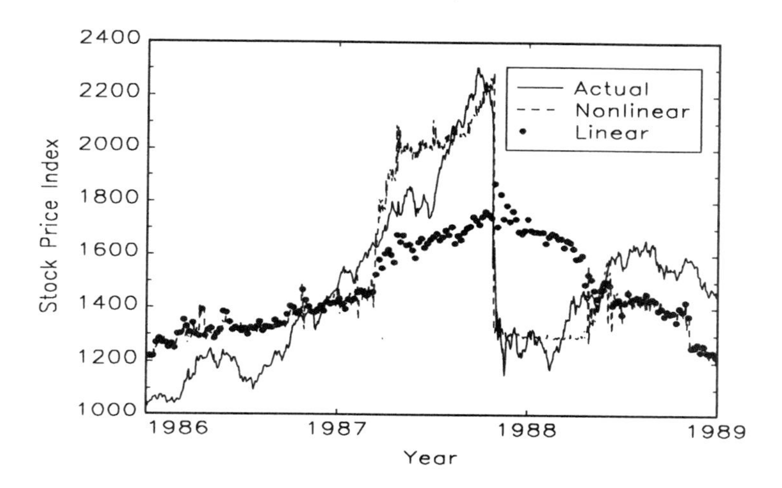

Figure 9.5: Comparison of actual with linear and nonlinear predictions

emphasising that this result is achieved without the aid of dummy variables and without an *ex post* inspection of the data. The nonlinear model also captures the general surge in the share price prior to the stock market crash.

It is interesting to note that the linear and nonlinear models yield similar predictions during the first and third periods, namely the periods of unimodality. The predictions of the two models diverge during the second period, the period of bimodality and the period with the interesting dynamics. Over the second period the linear model is not able to cope with the sharp climb in the share price followed by an even sharper and more dramatic fall on 20 October 1987.

These results suggest that the period of bimodality can be interpreted as a bubble. This has two advantages. First, it provides a formal definition of what a bubble is; namely, a period of bimodality when the market contains two views of the price. Secondly, it provides an historical estimate of the start and the end of the bubble period.

9.3.4 Forecasting

To highlight the forecasting properties of the model, the model is estimated over three subsamples, all ending just prior to the crash. This also provides a test of the robustness properties of the model, as the bimodality period identified in Figure 9.4 may be driven by the stock market crash. A strong test of the model is to see if it can predict a crash when a crash has not occurred in the estimation period.

Three subperiods are chosen corresponding to twelve, seven and two working days prior to the crash. For each subperiod, a one-step-ahead forecast of the distribution is computed. If the distribution is bimodal, the probability to the left of the antimode, that is the area around the lower mode, is calculated. To estimate the one-step-ahead density, from inspection of (9.12) to (9.15) it is necessary to have information on future dividends. This is obtained from the forecasts of estimating a tenth order autoregression in dividends.

For the three subperiods the distribution is found to be bimodal. An estimate of the probability that the stock price index will crash using information up to and including twelve days before the time when it does crash is equal to 0.363. When the sample is increased by five days, this probability falls slightly to 0.325, although it is still rather high. Increasing the sample by another five observations, thereby including information up to and including the Friday just prior to the crash, yields a slight increase in the probability of a crash occurring of 0.333. All of these results suggest that, prior to the crash, the model is predicting that there is a relatively high probability that the price will fall.

9.4 Conclusions

This chapter has provided a framework for modelling discrete jumping behaviour by building a flexible model which can exhibit multiple equilibria. This framework has several advantages. First, it represents a strategy for uncovering the reasons why processes display jumping behaviour. Secondly, it avoids the need for the *ad hoc* use of dummy variables. Thirdly, the timing of jumps is determined endogenously within the model.

Key features of the model have been demonstrated using simulation experiments. The framework was also applied to modelling price bubbles in the Australian stock market. The main result of the empirical analysis was that the stock market was operating in a zone of multiple equilibria around the time of the crash in October 1987 and that the large fall in the stock price was the result of the stock price jumping between equilibria. An important feature of this empirical analysis is that both the timing and magnitude of large jumps can be modelled endogenously with the aid of higher order moments without the need for dummy variables.

The analysis undertaken in this chapter has concentrated on modelling structural change when the timing of the break is unknown. One way the analysis can be extended is in constructing formal tests of structural change along the lines of Andrews (1993). Another extension is to relate the analysis to the work of Akgiray and Booth (1988) which models volatility in stochastic continuous time models with the imposition of a Poisson jump process.

Chapter 10

A Topological Test of Chaos

Eugene S.Y. Choo

Empirical attempts to test for chaotic behaviour in economic and financial data have tended to generate inconclusive results. The main tools used in testing for chaos are the correlation dimension of Grassberger and Proccacia (1983a, b), the Lyapunov exponents procedure of Guckenheimer (1982) and Wolf *et al.* (1985), and the BDS statistic of Brock *et al.* (1987). A potential problem with these methods is that the data sets need to be relatively large and free of noise. These properties are not satisfied by economic data, and although financial data sets tend to be large they none the less are, or at least appear to be, noisy.

Some modifications to these tools have been suggested to achieve robustness to noise and the smallness of the size of data sets. Ramsey and Yuan (1989, 1990) developed a procedure for reducing the bias in the correlation dimension, while McCaffrey *et al.* (1992) used a nonparametric regression procedure based on neural networks for estimating the dominant Lyapunov exponent.

The correlation dimension and Lyapunov exponents can be classified as metric approaches as they are concerned with measuring the distances between points on an attractor and computing the associated rates of convergence and expansion of these points. In contrast, the approach by Mindlin *et al.* (1990, 1991) is based on the topological properties of chaotic processes. The procedure involves the use of a close returns plot to test for the presence of chaotic behaviour. This procedure was shown to apply well to small data sets and was found to be robust to the presence of noise. It was also shown to be capable of providing information regarding the dynamics of the data set.

In another line of research, Eckmann *et al.* (1987) developed the recurrence plot which was modified by Koebbe and Mayer-Kress (1992) to a form almost similar to the close returns plot. An important difference between the close returns and recurrence plots is that the latter was developed for multivariate systems whereas the former was developed for a univariate system. In fact the two methods are equivalent for a univariate system. As the allowance for modelling multivariate systems in recurrence plots has been shown to be advantageous in uncovering underlying dynamics of processes from data, it is the aim of this chapter to generalise the close returns plots to multivariate systems. A key result of this chapter is that the multivariate extension of the close returns plot is found to be more effective in isolating segments of the data set that exhibit chaotic behaviour.

The close returns procedure is outlined in Section 10.1 while Section 10.2 contains a discussion of its properties. The properties of close returns plots for economic processes are examined in Section 10.3, with special attention given to processes exhibiting near unit roots and autoregressive conditional heteroscedasticity. Section 10.4 contains the results of computing close returns plots for a range of Australian data with special attention given to business cycles and financial markets.

10.1 The Topological Approach

The topological approach developed in this section for detecting nonlinear structure and possibly chaos is the multivariate analogue of the close returns plot initially developed by Mindlin *et al.* (1990, 1991). This procedure relies on the organisation of unstable periodic orbits that forms the strange attractor of a chaotic series. The period of an orbit is defined as the time, measured in units of the sampling interval, it takes for the orbit to be completed. This graphical procedure searches for these unstable periodic orbits in the time series while maintaining the time ordering of the data. This allows certain geometric features in the plot to be related to the underlying dynamics within the data.

Consider the unknown data generating process of an n dimensional system denoted by f, where $f : \Re^n \rightarrow \Re^n$and $\{x_t\}_{t=1}^{T}$ is an observed time ordered sequence of univariate data, that is $x_t \in x_t^i$ and $i = \{1, ..n\}$. The technique of phase space reconstruction allows us to generate a sequence of time-ordered vectors $\{\vec{x}_t\}_{t=1}^{Nv}$ with Nv denoting the total number of reconstructed vectors in m dimensional space. This is done by taking delayed or lagged values of the univariate time series x_t to form components of the reconstructed vector. The delay factor in terms of units of sampling interval is assumed to be 1 in the following analysis. For example, a reconstruction with an embedding of

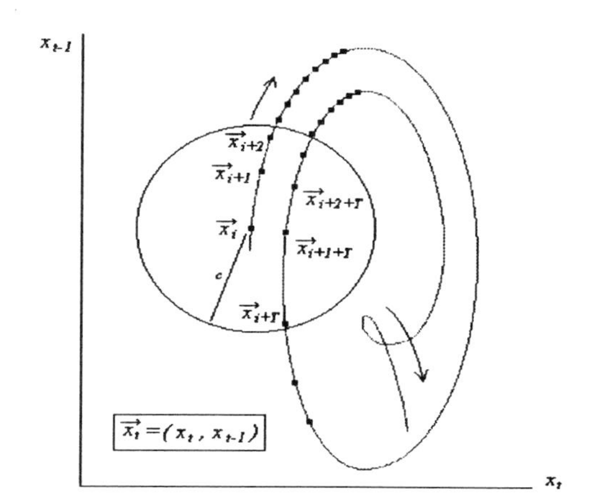

Figure 10.1: Close returns of reconstructed vectors in a strange attractor

6 and a delay factor of 1 generates the reconstructed vector:

$$\vec{x}_t = (x_t, x_{t-1}, ..., x_{t-4}, x_{t-5}) \tag{10.1}$$

From Takens Embedding Theorem, $\{\vec{x}_t\}_{t=1}^{Nv}$ creates an artificial system that can uncover the dynamics of f provided that the embedding dimension chosen is large enough, that is $m > 2n + 1$. If the trajectories of the solutions to the system are chaotic, the sequence of m dimension reconstructed vectors also navigate paths that describe a strange attractor.

Figure 10.1 illustrates the orbital paths of a sequence of reconstructed vectors about an attractor in two-dimensional space. Consider a reference vector $\vec{x}_i$, which lies within a periodic orbit of an attractor. Subsequent vectors, $\vec{x}_{i+1}, \vec{x}_{i+2}, \vec{x}_{i+3}, ...$, execute the orbital path for a number of sampling intervals before being repelled away from the attractor as shown in Figure 10.1. This orbital path is referred to as an unstable periodic orbit, where the reconstructed vectors only execute the orbital path for a short time before being repelled away from the space in which the trajectory is orbiting. The strange attractor of a chaotic series comprises a dense set of unstable orbits of different periodicity. It is these unstable periodic orbits that are captured by a close returns plot.

Within each periodic orbit, some vectors return to the neighbourhood of the reference vector $\vec{x}_i$after some sampling interval T. This is represented in Figure 10.1 by the vector $\vec{x}_{i+T}$. Let $\delta(i, i+T)$ denote the Euclidean distance between the reference vector $\vec{x}_i$ and $\vec{x}_{i+T}$:

$$\delta(i, i+T) = \left\| \vec{x}_i - \vec{x}_{i+T} \right\| \tag{10.2}$$

The vector $\vec{x}_{i+T}$ is said to be in close returns or recurrence to the reference vector $\vec{x}_i$if $\delta(i, i+T)$ is small, for example less than some tolerance c, represented by the circle centred around the reference vector. Within each periodic orbit, consecutive pairs of vectors, for example $\delta(i+1, i+1+T)$, $\delta(i+2, i+2+T)$, and so on, are also 'close' forming a region of close returns or recurrence.

This region of close returns can be mapped using a function $\rho : \Re^l \to \Re^k$ where k denotes the number of colours to be used in the plot. Each of these colours represent a specified degree of closeness. For example, a three-colour map has the function ρ, defined by equation:

$$\rho[\delta(i, i+T)] = \left\{ \begin{array}{lll} 0 & \text{if} & \delta(i, i+T) > c_1 \\ 1 & \text{if} & c_2 < \delta(i, i+T) \leq c_1 \\ 2 & \text{if} & c_3 < \delta(i, i+T) \leq c_2 \end{array} \right\} \quad \text{when } k = 3 \tag{10.3}$$

The close returns plots used in this chapter are confined to a two colour map of black and white. The choice of the cutoff values c used to identify regions of close returns is determined as:

$$c = \lambda \max \left\{ \left\| \vec{x}_i - \vec{x}_j \right\|, \text{ where } i, j \in \{1, ...Nv\} \text{ and } i \neq j \right\} \tag{10.4}$$

with λ varying from 0.01 to 0.6. If λ is set to a very small value, only a small number of vector pairs are 'close', hence generating only a small number of black points which makes it difficult to identify the underlying patterns that characterise the data. The converse happens if λ is set to a very large value, because this generates large clusters of black points as now the number of pairs of reconstructed vectors in close returns is substantially increased, possibly obscuring the underlying pattern..

To generate the close returns plot, all the Euclidean distances between the respective vectors, $\left\| \vec{x}_i - \vec{x}_{i+t} \right\|$, where $i = \{1, ..Nv\}$ and $t = \{1, ..Nv - 1\}$ are initially calculated. The close returns plot graphs the index i on the horizontal axis where $i = \{1, ..Nv\}$ against t, the sampling distance between vectors that are in close returns on the vertical axis where $t = \{1, .., Nv - 1\}$. If the Euclidean distance $\delta(i, i+t)$ is less than the cutoff value c, indicating close returns, a black point is plotted at (i, t) of the graph, and if it is greater than c, nothing is plotted. The close returns plot, in short, is a plot of the index of reconstructed vectors against the sampling time, t, when vectors are at close returns. Regions of close returns in the form of unstable periodic orbits in the data, as is characteristic of a strange attractor, are represented by segments of horizontal lines in the plot.[1]

10.2 Properties of Close Returns Plots

The key properties of close returns plots are summarised as follows:

1. *Periodic*: Horizontal lines running across the entire plot at vertical intervals determined be the period of the cycles.
2. *Quasi-periodic:* Contour-like patterns in between horizontal lines.
3. *Chaotic:* Scattered horizontal line segments.
4. *Stochastic:* Uniformly scattered points.

These properties are now discussed in turn with the aid of the Rossler attractor:

$$\begin{aligned}\frac{\partial x}{\partial t} &= -y - z \\ \frac{\partial y}{\partial t} &= x + \alpha y \\ \frac{\partial z}{\partial t} &= \beta - \gamma z + xz \qquad (10.5)\end{aligned}$$

where $\alpha, \beta, \gamma > 0$, are parameters. The key parameter that determines the trajectories of the variables x, y and z is γ. It determines the critical point where period doubling occurs and chaos emerges. The data used in this section are generated by discretising the system with a sampling time interval set at 0.03 seconds.

10.2.1 Periodic Behaviour

Setting the parameters in (10.5) at $\alpha = \beta = 0.2$ and $\gamma = 2.4$, the trajectories of the solutions are periodic. The close returns plots with the cut-off value $c = 0.07$ and embedding dimensions of $m = 1$ and $m = 7$ are shown in Figures

[1] A variation of the close returns procedure is the histogram of close returns introduced by Gilmore (1993). It is a histogram of the sum of incidence of close returns calculated at each sampling interval t.

10.2(a) and 10.2(b) respectively. The embedding is increased to examine the stability of the dynamics corresponding to the patterns observed in the plots of lower dimension. Note the equally spaced horizontal and negatively sloped lines in the two plots which are characteristics of data that is periodic. Consider a horizontal line segment of length l starting from a point (i_0, t_0). This line represents an array of points:

$$\left\{(i, \phi(i)) : i = i_0, i_{0+1}, i_{0+2}, \ldots.., i_{0+(l-1)} \text{ where } \phi(i) = t_0\right\} \tag{10.6}$$

or:

$$\{(i, t), (i+1, t), (i+2, t), \ldots\ldots (i+l-1, t)\} \tag{10.7}$$

which correspond to the distances:

$$\left\{\left\|\vec{x_i} - \vec{x_{i+t}}\right\|, \left\|\vec{x_{i+1}} - \vec{x_{i+1+t}}\right\|, \ldots, \left\|\vec{x_{i+(l-1)}} - \vec{x_{i+(l-1)+t}}\right\|\right\} \tag{10.8}$$

This suggests that the horizontal lines correspond to consecutive pairs of equally spaced reconstructed vectors that are close to one another and remain close for an interval of time. This region of close returns occurs when the sequence of vectors is constructed from data points coming from trajectories of the same periodic orbit.

If these horizontal lines traverse the entire plot at evenly spaced vertical intervals, it suggests that the periodicity of the orbit is constant and stable throughout the entire data set. In Figures 10.2(a) and 10.2(b), the horizontal lines appear after every 185 or so vertical intervals, suggesting that the sequence of reconstructed vectors returns to the neighbourhood of the reference vector after 185 or so units of sampling interval. This is true for all subsequent reference vectors in this periodic data set, resulting in the complete horizontal lines observed.

Figure 10.2(a) shows evenly spaced lines with a negative slope of -2. The array of points describing a line of slope, say g and length l, passing through (i_0, t_0) is given by:

$$\left\{(i, \phi(i)) : i = i_0, i_{0+1}, \ldots, i_{0+(l-1)};\ \phi(i) = t_0 + g(i - i_0)\right\} \tag{10.9}$$

If the gradient is set at -2, then the array of coordinates becomes:

$$\{(i, t), (i+1, t-2), (i+2, t-4), \ldots\ldots (i+l-1, t-2(l-1))\} \tag{10.10}$$

corresponding to distances:

$$\left\{\left\|\vec{x_i} - \vec{x_{i+t}}\right\|, \left\|\vec{x_{i+1}} - \vec{x_{i+t-l}}\right\|, \ldots, \left\|\vec{x_{i+(l-1)}} - \vec{x_{i+t-(l-1)}}\right\|\right\} \tag{10.11}$$

The spacing in terms of units of sampling interval between pairs of reconstructed vectors that are in close returns are gradually decreasing. Such

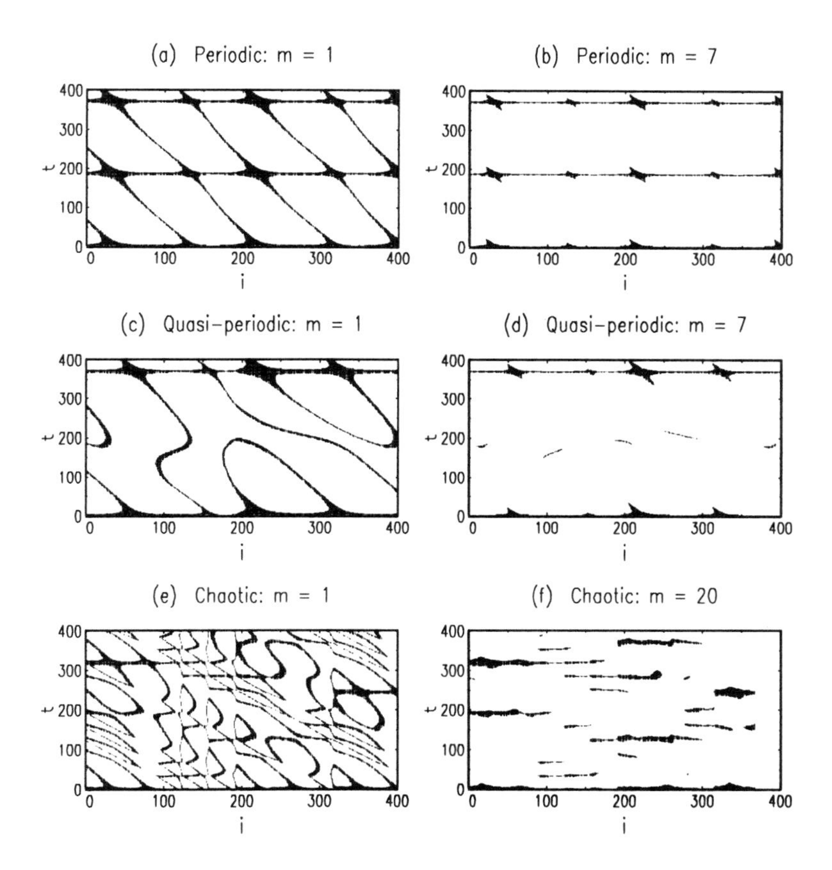

Figure 10.2: Close returns plots of the Rossler attractor

a pattern occurs when a sequence of reconstructed vectors are inverted in time elsewhere in the data set; see Koebbe and Mayer-Kress (1992). In the present case where the data are periodic, the trajectory follows a sinusoidal time path that is symmetrical. Here a sequence of data has identical values in an inverse order when completing a periodic cycle. The negatively sloped lines become less visible when the embedding dimension is increased as shown in 10.2(b). This suggest that the inversion pattern present in a vector sequence of lower embedding are gradually lost when the dimensional embedding is increased.

Increasing the dimensional embedding introduces different data values as additional components of each reconstructed vector. If the new data values are about the same magnitude as components of the original pair of vectors, then the close returns are expected to be maintained between the initial pair of reconstructed vectors as dimensional embedding increases. This is likely to be the case when components of the reconstructed vectors are taken from periodic data. The horizontal line patterns which represent close returns between equally spaced reconstructed vectors from the same periodic orbit remain as the dimensional embedding increases.

10.2.2 Quasi-Periodic Behaviour

Increasing the parameter γ of the Rossler system in (10.5) from $\gamma = 2.4$ to $\gamma = 4.0$ causes period doubling to occur. The data are now aperiodic or quasi-periodic, comprising trajectories of two or more periods. No longer are the orderly linear patterns associated with periodic data observed. Instead the negatively sloped lines have merged to form contour-like patterns, as shown in Figure 10.2(c). Increasing the dimensional embedding to $m = 7$, as in Figure 10.2(d), shows that these contour-like patterns fade away leaving behind the horizontal line segments traversing the entire plot. In this scenario, additional data introduced as components of the reconstructed vector as dimensional embedding increases are less likely to come from the same periodic orbit. The increased variability in the data is responsible for the more complex patterns observed in the plots. A detailed analysis of the actual dynamics associated with the specific contour-like patterns is not pursued here. Suffice it to say that, as the data get more chaotic, plot patterns become progressively more complicated.

10.2.3 Chaotic Behaviour

The Rossler system in (10.5) generates chaotic trajectories when $\alpha = 0.343$, $\beta = 1.82$ and $\gamma \doteq 8$. Close returns plots are shown in Figures 10.2(e) to 10.2(f) for embedding dimensions $m = 1$ and $m = 20$ respectively. The value of $\lambda = 0.12$ in (10.4). A more dense contour-like pattern is obtained in the plot of low-dimensional embedding. The pattern comprises mainly a combination of merged negatively sloped lines as well as horizontal line segments. Like the quasi-periodic case, the contour-like pattern gradually fades away as the dimensional embedding is increased from $m = 1$ to $m = 20$, leaving behind segments of horizontal lines scattered throughout the entire plot.

Figure 10.2(f) displays line segments that are vertical and those with a negative slope of -1. These arise when a vector $\vec{x}_i$ is in close returns with a sequence of successive reconstructed vectors. For a chaotic system, this happens when the vector executes an orbital path centred at a small geometric region in phase space such that a sequence of successive vectors is never too far from the reference vector. These dynamics are also responsible for the line segments with the slope of -1 which map the region of close returns when a vector is close to a sequence of preceding reconstructed vectors.

When the data become chaotic, increasing the dimensional embedding is likely to introduce, as additional components of a reconstructed vector, data from orbits of different periodicity. These are more likely to differ in magnitude from components of the original vector. Since close returns are most likely to occur between pairs of vectors whose components come from the same periodic orbit, we expect only segments of horizontal lines to remain as the embedding is increased. These represent the unstable periodic orbits which characterise a strange attractor. The introduction of dimensional embedding to the close returns procedure has the advantage of isolating these horizontal line segments. This is a feature that the initial version of the close returns procedure, restricted only to scalar time series, lacks.

10.2.4 Stochastic Behaviour

To investigate the effect of noise on the close returns plot, the chaotic series x_t from the Rossler system in (10.5) is adjusted as:

$$x_t^* = x_t + \mu \sigma_x \varepsilon_t \tag{10.12}$$

where x_t^* is the newly generated series, σ_x is the standard deviation x_t, $\varepsilon_t \sim N(0, 1)$, and μ controls the level of noise input.

The effect of increasing μ on the close returns plots is to make the underlying patterns fuzzy. In the limit as $\mu \to \infty$, x_t^* is dominated by noise and the close returns plot just displays a random scattering of points.

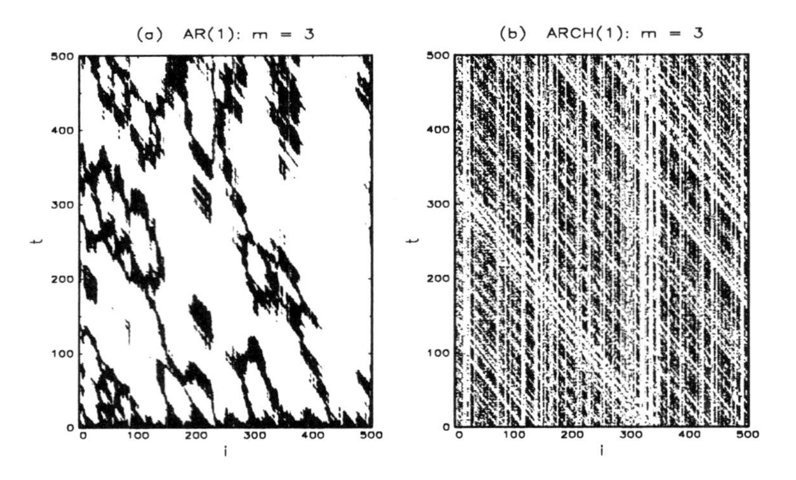

Figure 10.3: Simulated economic models

10.3 Simulated Economic Models

10.3.1 Near Unit Processes

The metric procedures like the correlation dimension procedure are unable to discriminate between a unit root, or near unit root, and a chaotic process; see Brock and Sayers (1988). To investigate the ability of the close returns plot, Figure 10.3(a) contains the results from simulating the following near unit root process with an embedding dimension of $m = 3$:

$$y_t = \alpha_0 + \alpha_1 y_{t-1} + \varepsilon_t \tag{10.13}$$

where $\alpha_0 = 0$ and $\alpha_1 = 0.98$ and ε_t iid $N(0, 1)$. Inspection of this plot shows that it does not resemble any of those presented above from a chaotic, periodic or quasi-periodic process. Nor does it resemble a purely random process. Dense clusters of points are observed which form random diamond-like patterns throughout the entire plot. There are also line segments that are negatively sloped, suggesting possible inversion as discussed earlier. The orientation of points suggests strongly that there is an underlying structure within the system.

10.3.2 ARCH

The procedure is now applied to processes exhibiting nonlinear dependence in the form of autoregressive conditional heteroscedasticity (ARCH) error. This is done to investigate the ability of the close returns procedure to distinguish between a chaotic process and nonlinear dependence. The metric procedures such as the BDS statistics is unable to differentiate between linear, nonlinear, stochastic or even deterministic forms of dependence.

The model is generated as:

$$\begin{aligned} y_t &= \varepsilon_t h_t^{0.5} \\ h_t &= \beta_0 + \beta_1 \varepsilon_{t-1}^2 \end{aligned} \tag{10.14}$$

where ε_t iid $N(0,1)$. The close returns plot with an embedding dimension of $m = 3$, is shown in Figure 10.3(b). Unlike the autoregressive process given in 10.3(a), there is no clear obvious pattern in the plot. The points are scattered into vertical and diagonal line paths, unlike the clusters observed when the process is autoregressive. The scattering also lacks the randomness that would be observed for a purely random process.

10.4 Application to Australian Data

10.4.1 Business Cycles

An important question in analysing business cycles concerns whether the cycles are endogenously determined within the economic system. Evidence of chaotic behaviour in these series provides support for the existence of an endogenous propagation system and suggests that the equilibrium fluctuations in the form of business cycles are not necessarily the product of exogenous shocks. The existence of an endogenous deterministic system also suggests that the government should engage in stabilisation policies.

A number of studies using the standard metric approaches have tested business cycles for nonlinear dependencies and low dimensional chaos; see, for example, Brock and Sayers (1988) who found possible evidence of chaos in the US business cycle, and Frank and Stengos (1988) who found no such evidence for the Canadian business cycle. To investigate the properties of the Australian business cycle, Figure 10.4 gives the close returns plot for Australian real GDP for embedding dimensions $m = 1, 5$. The data are quarterly, begining in September 1959 and ending in June 1994, a total of 140 observations, and are seasonally adjusted and detrended using a linear trend.

The close returns plots in Figure 10.4 show no obvious horizontal line pattern that characterises chaotic, periodic or quasi-periodic processes. The

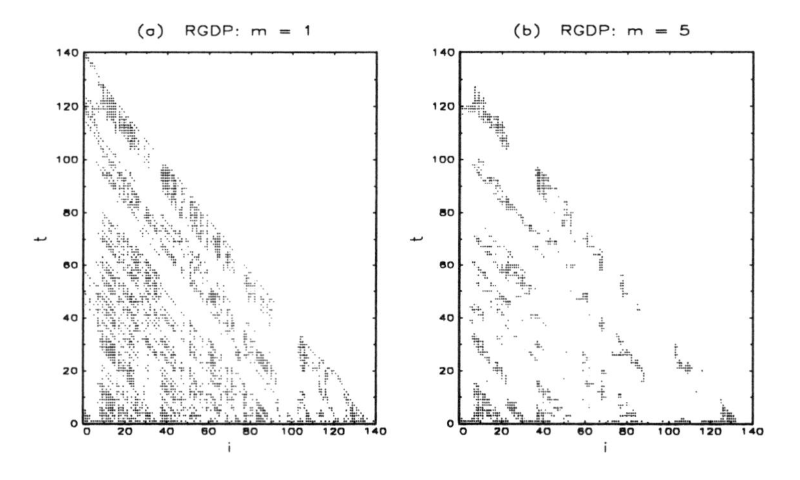

Figure 10.4: Close returns plots of the Australian business cycle

patterns however suggest that the fluctuations around the trend are in no way random.[2] Although it is not possible to identify explicitly the underlying structure of the cycle, it is possible to rule out evidence of chaotic behaviour and conclude that the equilibrium fluctuations around the trend are not a product of an endogenous deterministic system. The study by Gilmore (1993) on the US business cycle yields the same qualitative results.

10.4.2 Financial Markets

Theory suggests that if financial markets are frictionless and functioning perfectly, changes in asset prices should be unpredictable. Chaos theory, on the other hand, suggests that the trajectories of a chaotic system, although having the appearance of being random, are potentially perfectly predictable; see Brock and Malliaris (1989, p.325). If the efficient market theory is correct, there should be no evidence of chaos in asset price changes.

There is some evidence based on metric approaches that financial data exhibit chaotic behaviour; see for example LeBaron and Scheinkman's (1987) study of daily stock returns data, and the analysis of US. Treasury Bill rates

[2]Similar results were obtained using other Australian macroeconomic data, but are not reported here.

by Brock and Malliaris (1989). To investigate the presence of chaos in Australian financial data close returns plots are constructed for three Australian daily financial returns: share returns, gold returns and foreign returns. All returns are computed as daily differences in prices. The share price index and gold price index start on 1 July 1990 and end on 29 June 1994, a total of 1014 observations. The exchange rate is Australian/US which starts on 3 January 1984 and ends on 29 June 1991, a total of 1885 observations. All data used are obtained from the Reserve Bank of Australia and are available on request from the author.

The close returns plots for the three financial series are given in Figure 10.5 for embedding dimensions of $m = 1, 3$. All plots display similar characteristics. In particular, the orientation of points into vertical and diagonal line paths suggests that the inherent structures in the three returns are not random. There is, however, no evidence of chaos as would occur if horizontal line segments were present. A failure to find evidence of chaos in Australian stock returns and interest rates is consistent with the results of Gilmore (1993) for US stock returns. The presence of structure in foreign returns and a lack of evidence of chaos is consistent with the results of Hsieh (1989a) who found evidence of nonlinear dependence in daily exchange rates but was not able to determine whether the dependence was the result of chaotic behaviour.

10.5 Conclusions

Procedures for the detection of chaotic behaviour in time series data have been restricted to standard metric methodologies based on correlation dimensions, Lyapunov exponents and BDS statistics. These procedures do not apply well to small and noisy data that are often encountered in economics, thereby making conclusions based on these procedures tenuous. The close returns procedure, a topological approach of detecting for chaotic behaviour, offers a promising alternative which can overcome some of the difficulties assocated with metric aproaches. In particular, the method was shown to be robust to reasonably high levels of noise and small data size. The general topological approach also has the potential to quantitatively reconstruct a system characterised by chaos.

The close returns plot was applied to testing for evidence of chaotic behaviour in the Australian business cycle and Australian financial markets. The key result was that there was no evidence of chaos. There was, however, evidence of structure. While it was not possible to identify the explicit form of this structure using the close returns plots, this represents a fruitful direction for future research.

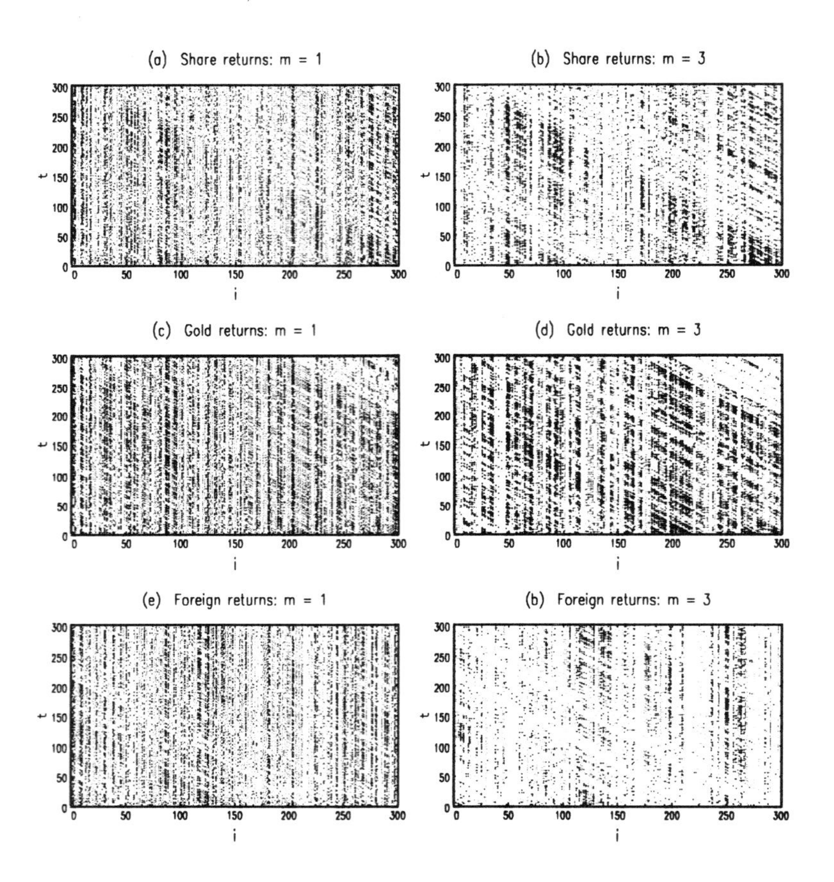

Figure 10.5: Close returns plots of Australian financial series

Chapter 11

Genetic Algorithms and Trading Rules

Robert Pereira

This chapter uses a genetic algorithm to search for optimal parameter values for some simple popular technical trading rules, namely the channel, filter and moving average rules. Genetic algorithms are biologically inspired optimisation techniques which are efficient at searching large solution spaces. By using genetic operations they are able to simulate an evolutionary process by which an initial population of potential solutions to a problem can develop through successive generations into optimal solutions. Due to their speed and flexibility, genetic algorithms are well suited for the development of trading rules for real time trading; see Allen and Karjalainen (1994), Bauer (1994) and Deboeck (1994).

Genetic algorithms are important in devising trading rules for the following reasons. First, the trading rule parameter values chosen by financial market traders are usually determined in an *ad hoc* subjective manner based largely on experience. A more rational approach would be to search for the values that yield the greatest profits using historical data.[1] These optimal parameter values can be found using either an exhaustive grid-search procedure or traditional gradient-type algorithms, such as the Newton and quadratic hill-climbing techniques. However, an exhaustive grid search procedure is very time consuming when the potential solution space is large,

[1]Lukac and Brorsen (1989) conduct a study of the usefulness of historical data in selecting parameters for technical trading rules. They conclude that the forecasting ability of optimisation based on historical data is limited. However, these results are based on only two trading rules, the channel and directional movement rules, using futures market data. Therefore, their results cannot be generalised to all financial markets and rules.

while gradient-type algorithms often only locate local optima when the solution space is highly nonlinear, displaying sharp peaks. Dorsey and Mayer (1995) examine the ability of a genetic algorithm compared with traditional search methods, to solving optimisation problems that display multiple optima, nondifferentiability and other irregular features. They conclude that, although the genetic algorithm does not always find the global optimum, it arrives at a solution that is closer to the global optimum compared to the other methods which are more likely to get anchored at local optima.

According to Pruitt *et al.* (1992) technical traders usually do not make decisions based on a single technical indicator.[2] Therefore it may be beneficial to develop complex combinations of simple trading rules. Bauer (1994) employs a genetic algorithm to devise market timing strategies for the US stock and bond markets that are based on fundamental analysis. Allen and Karjalainen (1994) use genetic programming, a slight variation of a genetic algorithm, to develop profitable complex combinations of technical trading rules for the stock market and foreign exchange market respectively.

Also, as the behaviour of financial market prices can fundamentally change over time, trading rules and strategies should remain flexible and thus be periodically updated to reflect structural changes in markets. Genetic algorithms can be employed to search efficiently for the optimal parameter values, develop optimal combinations of simple trading rules, and determine the optimal frequency with which rules should be revised due to changes in fundamentals. This can be done relatively more quickly and accurately with genetic algorithms than with other methods, which is important for real time trading in financial markets.

This chapter considers the first issue, using a genetic algorithm to select the optimal values for trading rule parameters. A genetic algorithm is developed and then used to search for the optimal parameter values, based on historical data on the All Ordinaries Index, for each of the technical trading rules considered in this study. The profitability of these rules, employing the optimal parameter estimates, is then tested using an out-of-sample test period.

Section 11.1 discusses alternative technical trading rule. Section 11.2 outlines the development of a genetic algorithm used for searching the optimal trading rule parameters. The out-of-sample profitability and statistical significance of these rules applied to the Australian share market, employing the derived parameter estimates, are evaluated in Section 11.3.

---

[2] Technical traders are financial market traders who base their trading decisions on the investment philosophy of technical analysis.

11.1 Technical Trading Rules

Trading rules are used by financial market traders to assist them in determining their investment position in a particular financial asset. Technical trading rules are based on technical analysis which assumes that prices slowly reflect information. By using historical price and volume data, these rules attempt to profit by predicting the direction of future price movements. Some of the more popular rules currently employed by traders are based on technical indicators such as filters, moving averages, relative strength indices, and support and resistance levels; see Eng (1988).

Trading rules return either a buy or a sell signal which determines the investment position taken in a particular security. If the current investment position is short, then a buy signal results in purchasing the security while a sell signal leads to staying short. If the current investment position is long, a buy signal results in remaining long while a sell signal leads to selling the security. In this study five technical trading rules are considered: these are the simple moving average, filtered moving average, double moving average, filter, and channel rules.

11.1.1 Moving Average Rules

Moving averages are widely used by traders to identify trends in financial market prices. A moving average (MA) is simply an average of current and past prices over a specified period. An MA of length L is calculated as:

$$MA_t(L) = \frac{1}{L}\sum_{i=0}^{L-1} P_{t-i}$$

for all values of L greater than zero. By smoothing out the short-term fluctuations or noise in the price series, the MA is able to capture the underlying trend over a particular period of historical prices. However an MA can only indicate that a trend has begun *ex post* since it is based on current and past prices. In a trending market this doesn't present a serious problem because by definition in a trending market, trends persist for some time. An MA can be used to formulate a trading rule to determine whether to buy or sell a security.

With the simple MA rule, if the price rises above the L day MA, then the security is bought and held until the price falls below the L day MA at which time the security is sold. This simple rule can be modified to create the filtered MA rule and the double MA rule. A filtered MA rule is similar to the previous rule except that it includes a parameter, known as a filter, which accounts for the percentage increase or decrease of the price relative to its MA. The inclusion of this filter helps reduces the number of false buy and sell signals created by a simple MA rule which can occur when the movement of the market price over time is sideways instead of trending either up or down. This rule operates by returning a buy signal if the price rises by X per cent above the L day MA and then returning a sell signal only when the price falls by X per cent below the L day MA at which time the security is sold. In contrast to the previous two rules, a double MA rule compares two MAs of different lengths. With this rule if the shorter (S) length MA rises above the longer (L) length MA from below then the security is bought and held until the shorter MA falls below the longer MA at which time the security is sold.

The three different MA rules can be nested in the following decision rule:

$$\frac{1}{S}\sum_{i=0}^{S-1} P_{t-i} - \frac{(1+(1-2d_t)\,X)}{L}\sum_{j=0}^{L-1} P_{t-j} \tag{11.1}$$

$$= (MA_t(S) - (1+(1-2d_t)\,X)MA_t(L))\left\{\begin{array}{l} >0, Buy \\ \leq 0, Sell \end{array}\right. \tag{11.2}$$

where S and L are the lengths of the short and long MAs respectively and X is the filter parameter, while d_t is a dummy variable defined as:

$$d_t = \left\{\begin{array}{l} 1, Buy \\ 0, Sell \end{array}\right. \tag{11.3}$$

where d_t takes the value of 1 (0) if the rule returned a buy (sell) signal in the previous period.

The respective rules can be applied by imposing the following restrictions:

1. *Simple MA:* $L > 0$, $S = 1$ and $X = 0$. Equation (11.1) reduces to:

$$P_t - \frac{1}{L}\sum_{j=0}^{L-1} P_{t-j} = (P_t - MA_t(L))\left\{\begin{array}{l} >0, Buy \\ \leq 0, Sell \end{array}\right.$$

2. *Filtered MA*: $L > 0$, $S = 1$ and $X \geq 0$. Equation (11.1) becomes:

$$P_t - \frac{(1+(1-2d_t)\,X)}{L}\sum_{j=0}^{L-1} P_{t-j}$$

$$= (P_t - (1+(1-2d_t)\,X)MA_t(L))\left\{\begin{array}{l} >0, Buy \\ \leq 0, Sell \end{array}\right.$$

3. *Double MA:* $0 < S < L$ and $X = 0$. Equation (11.1) reduces to:

$$\frac{1}{S}\sum_{i=0}^{S-1} P_{t-i} - \frac{1}{L}\sum_{j=0}^{L-1} P_{t-j} = (MA_t(S) - MA_t(L)) \left\{ \begin{array}{l} > 0, Buy \\ \leq 0, Sell \end{array} \right.$$

11.1.2 Rules Based on Order Statistics

Channel and filter rules are based on the idea that when financial market prices rise above (drop below) a certain level they will continue to rise (fall) for some period of time. This assumes that there are certain periods of time when prices are autocorrelated and these rules attempt to profit from identifying these periods. Order statistics, such as the maximum and minimum of the historical price series, are used to identify these levels in prices. The current maximum price $P_t^{\max}(L)$ and the minimum price $P_t^{\min}(L)$ for a historical price series of length L are:

$$P_t^{\max}(L) = Max\left[P_{t-1}, ..., P_{t-L}\right]$$

$$P_t^{\min}(L) = Min\left[P_{t-1}, ..., P_{t-L}\right]$$

respectively.

The channel (or trading range break) rule is founded on the idea of support and resistance levels which are related to the market forces of demand and supply. The support level is achieved at a price where buying power dominates selling pressure, effectively placing a floor on the level of prices. The resistance level, the opposite to support, is defined as the price where selling pressure exceeds buying power forcing down the price and effectively creating an upper level or ceiling in prices. With the channel rule the resistance (support) level is defined using the maximum (minimum) price over the previous L days. This rule returns a buy (sell) signal when the price of the security breaks through the current resistance level or ceiling (support level or floor).

The filter rule is based on the idea that when price increases by X per cent, represented by the filter, above a previous low, the security should be bought and held until the price falls by X per cent below a previous high. The previous high (low) is given by the local maximum (minimum) price determined by the maximum or minimum price of the historical price series commencing since the last transaction.

These two rules can be nested in the following decision rule:

$$\left(P_t - (1 + (1 - 2d_t)\,X)\left(P_t^{\max}(L)\right)^a \left(P_t^{\min}(L)\right)^{(1-a)}\right) \left\{ \begin{array}{l} > 0, Buy \\ \leq 0, Sell \end{array} \right. \quad (11.4)$$

where d_t is a dummy variable as defined in equation (11.3), X is the filter parameter and L is the length of the historical price series used in determining either the maximum or minimum price. While a, takes the value of d_t for the filter rule and $(1 - d_t)$ for the channel rule.

The *channel rule* can be derived by imposing the restrictions $a = 1 - d_t$, $L > 0$ and $X = 0$ in equation (11.4):

$$\left(P_t - (P_t^{\max}(L))^{(1-dt)} \left(P_t^{\min}(L)\right)^{d_t}\right) \begin{cases} > 0, Buy \\ \leq 0, Sell \end{cases}$$

The *filter rule* can be derived by imposing the restrictions $a = d_t, L =$ the number of days since the last transaction and $X \geq 0$ in equation (11.4):

$$\left(P_t - (1 + nX) \left(P_t^{\max}(L)\right)^{d_t} \left(P_t^{\min}(L)\right)^{(1-d_t)}\right) \begin{cases} > 0, Buy \\ \leq 0, Sell \end{cases}$$

11.2 Optimisation Procedures

A genetic algorithm can be used to search for the optimal parameter values for each of the trading rules:

1. *Simple MA:* L^*
2. *Filtered MA:* L^* and X^*
3. *Double MA:* L^* and S^*
4. *Channel rule:* L^*
5. *Filter rule*: X^*.

GAs, developed by Holland (1975), are a class of adaptive search and optimisation techniques based on natural evolution. By representing solutions to a problem using vectors of binary digits, referred to as individuals, mathematical operations, such as crossover and mutation, can be performed which are analogous to the genetic recombinations of the chromosomes of living organisms. New generations of individuals, the binary representations of the potential solutions, can be created and evolved over time using the crossover and mutation operators. However, as in nature, the fitter individuals have a better than average probability of surviving and reproducing relative to the less-fit individuals who eventually get eliminated. By incorporating a problem-specific fitness or performance function, an objective function, individuals' fitness can be assessed. A selection process based upon performance is applied to determine which individuals should participate in crossover, and thereby pass on their favourable traits to future generations. This evolutionary process continues until the fittest individuals, consisting of hopefully the optimal or near optimal solutions, dominate the population; for a detailed description see Goldberg (1989).

11.2.1 Problem Representation

To be able to solve the problem of finding the optimal parameter values for the rules considered here, potential solutions to this problem must be represented using vectors of binary digits. These vectors are linear combinations of 0s and 1s, for example [0 1 0 0 1]. Each of these binary representations is based on the binary number system which has an equivalent decimal value. For example, the decimal equivalent of the vector [0 1 0 0 1] $= (0 \times 2^4) + (1 \times 2^3) + (0 \times 2^2) + (0 \times 2^1) + (1 \times 2^0) = 8 + 1 = 9$.

Each type of rule has a different range of parameter values which results in a different number of elements in each of the vectors used as the binary representations.

Simple Moving Average Rule

Let the length of the simple MA rule (L) range between and include the integer values of 1 and 500. These values can be represented by their binary equivalents. The length of the binary representation (vector) in this case must ensure a maximum value of 500. A row vector with nine elements has a maximum decimal equivalent value of 511; [1 1 1 1 1 1 1 1 1] $= (1 \times 2^8) + (1 \times 2^7) + (1 \times 2^6) + (1 \times 2^5) + (1 \times 2^4) + (1 \times 2^3) + (1 \times 2^2) + (1 \times 2^1) + (1 \times 2^0) = 256 + 128 + 64 + 32 + 16 + 8 + 4 + 2 + 1 = 511$. While for a vector containing eight elements the maximum decimal equivalent value is only 255. Therefore a nine-element vector is chosen and its equivalent decimal value is restricted such that $1 \leq \mathrm{L} \leq 500$.

Moving Average With Filter

The filtered MA rule has two parameters L and X such that, $1 \leq L \leq 500$ and $0.00\% \leq X \leq 10\%$. The maximum number of different possible values of X are 1,001. To represent this range of values, a ten-element vector is necessary. This results in a maximum decimal equivalent value of 1023, corresponding to a vector of ten 1s, and the decimal equivalent values are restricted such that $0 \leq X \leq 1000$. To transform the decimal values into percentages simply requires dividing by 100. For example, $500 = 500/100 = 5$ per cent. While the parameter L can be represented by a nine-element vector, as described above for the case of the simple MA.

These two parameter values can be represented simultaneously by a vector containing nineteen elements, where the first nine elements correspond to the binary equivalent value of L and the last ten elements correspond to the binary equivalent value of X. This rule has a maximum number of 500,500 (500×1001) different possible values.

Double Moving Average

The double moving average rule has two parameters L and S, which both individually have the same range of values as the simple moving average, i.e. $1 \leq L \leq 500$ and $1 \leq S \leq 500$. In this case, however, a vector with eighteen elements is necessary. The first nine elements correspond to the length of one MA, while the last nine elements correspond to the length of the other MA. The shorter MA, which has length S, takes the lower value and the longer MA, which has length L, assumes the higher value. Since each MA has 500 different possible values, there exists a total of $500 \times 500 = 250{,}000$ different possible values for the double moving average rule.

Channel Rule

The parameter on the channel rule (L) is restricted to the values $1 \leq L \leq 500$. Similar to the simple MA rule, a nine-element vector is used.

Filter Rule

The parameter on the filter rule (X) can assume values from 0 per cent to 10 per cent. Similar to the filtered MA, the vector has ten elements and the decimal equivalent values are restricted to the desired range, $0 \leq X \leq 1000$.

11.2.2 Objective Function

Ultimately the goal of the genetic algorithm is to find the combination of binary digits referred to as the individual, representing a particular solution to the problem, which maximises the objective function. Therefore, each individual's performance or fitness can be assessed using this function. In the development of successful trading rules the goal is to search for the rule which maximises profits.

11.2.3 Genetic Algorithm Operations

Selection, crossover and mutation are the three important operations. It is through these mathematical operations that an initial random population of potential solutions to a problem is developed through successive generations into a final population consisting of a potentially optimal solution. The search process which ensues is highly effective because of these operations.

Selection involves determining which of the individuals in the current population will become the parents of the next generation. The genitor selection method, which is a ranking-based procedure developed by Whitley (1989), is used here. This approach involves ranking all individuals according to fitness and then replacing the worst performing individuals by copies of the better performing individuals. In the genetic algorithm developed in this chapter a copy of the best individual replaces the worst individual.

The method by which promising (relatively fitter) individuals are combined, is through a process of genetic recombination known as crossover. This ensures that the search process is not random but consciously directed into promising regions of the solution space. As with selection there are a number of variations; however, single point crossover is the most commonly used version and the one used in this study.

Crossover involves a number of steps. First, two individuals are randomly chosen from the current restricted population, with the least-fit individual deleted, to participate in crossover. Next a partitioning point at a particular position in the binary vector representation of each parent is randomly selected. At this break point the two vectors are partitioned, separating each vector into two subvectors. The right-hand side subvectors are exchanged between the two vectors and the finally the vectors are unpartitioned.

To illustrate the process of crossover assume that two vectors:

$$A = [1\ 0\ 1\ 0\ 0] \qquad B = [0\ 1\ 0\ 1\ 0]$$

are chosen at random and that the position of partitioning is randomly chosen to be between the second and third elements of each vector. Vectors A and B can be represented as $[1\ 0 : 1\ 0\ 0] = [A_1\ A_2]$ and $[0\ 1 : 0\ 1\ 0] = [B_1\ B_2]$ respectively, in terms of their subvectors. Whether or not crossover occurs is also determined randomly; given that crossover occurs then the subvector A_2 from vector A is switched with the subvector B_2 from vector B. Finally, both vector A and B are unpartitioned, yielding two new vectors or individuals $C = [1\ 0\ 0\ 1\ 0]$ and $D = [0\ 1\ 1\ 0\ 0]$.

New genetic material can be introduced into the population through mutation. This increases the genetic diversity in the population and, unlike crossover, randomly redirects the search procedure into new areas of the solution space which may or may not be beneficial. This action underpins the ability to find novel inconspicuous solutions and to avoid being anchored at local optimum solutions.

Mutation occurs by randomly selecting a particular element in a particular vector. If the element is a 1 it is mutated (switched) to 0. Otherwise, if it is a 0 it is mutated to a 1. This occurs with a very low probability, usually one element in a thousand, which is adopted in this study. Assuming that no other elements are selected for mutation except for the third in vector $C = [10010]$ then mutation would change this vector to $E = [10110]$.

11.2.4 The Steps Involved in a Genetic Algorithm

The first step is problem representation. This involves representing potential solutions to the problem as vectors consisting of binary digits. This procedure is discussed above in subsection 11.2.1. Once this is achieved an initial population of individuals is randomly created. Then the performance of each individual is evaluated using the objective function.

The next step, selection, involves deciding which individuals should be chosen to participate in the genetic recombination processes of crossover and mutation. The individuals of the initial population are ranked according to performance. Then the vector representing the worst individual is replaced by a copy of the best individual. The operations of crossover and mutation are then applied to the individuals in this current restricted population. These operations, which recombine and randomly change the elements of the vectors, lead to the creation of a generation of new individuals. Next, the performance of the individuals from this new generation is assessed using the objective function.

The last step involves checking a well-defined termination criterion. If this criterion is not satisfied, the genetic algorithm returns to the selection, crossover and mutation operations to develop further generations until this criterion is met, at which time the process of creation of new generations is terminated. The termination criterion adopted in this study is satisfied when either the population converges to a unique individual, which hopefully corresponds to the optimal solution, or when the maximum number of generations ($g^{\max}$) is reached. This latter condition ensures that the process does not continue indefinitely, which could occur in cases where the problem is such that the genetic algorithm does not converge. These steps are illustrated in Table 11.1.

11.3 Share Market Application

11.3.1 The Data

This study considers an application of the methodology discussed above to the Australian share market. The data, from the Equinet database, consist of the daily closing price of the All Ordinaries index and the daily 90-day Reserve Bank of Australia (RBA) bill dealer rate. The entire sample covers a

Table 11.1: The steps involved in one trial

Step	Action
1	Create an initial population of individuals randomly
2	Calculate the fitness of the individuals in the initial population
3	Replace the worst individual with a copy of the best (selection)
4	Apply genetic operators (crossover and mutation)
5	Calculate the fitness of the individuals in the new generation
6	Go to step 3 unless convergence or $g^{\max}$ is reached

period of approximately fourteen years, from 5 January 1982 to 30 November 1995, comprising 3628 observations. To avoid the problem of overfitting the data, the whole period is split into two distinct periods. These two periods are referred to as the rule-building period and the out-of-sample test period. The rule-building period is used to discover the optimal parameter values for the trading rules. It contains 2084 observations spanning approximately eight years, 5 January 1982 to 29 December 1989. The rules incorporating the parameter estimates chosen during this period are then tested out-of-sample in the testing period, 1 January 1990 to 31 November 1995, consisting of 1544 observations.

11.3.2 Descriptive Statistics

Table 11.2 provides descriptive statistics for the continuously compounded daily returns, $r_t = \ln\left(\frac{P_t}{P_{t-1}}\right)$, on the Australian share market as measured by the All Ordinaries index (P_t), and all returns are expressed as percentages. Daily returns are slightly positive for the Australian share market; ths is not surprising due to the long-run tendency of the share market price to increase over time. This can been seen in the plot of the All Ordinaries index in Figure 11.1. The October 1987 share market crash and the mini-crash of 1989 are evident in the time series plot of the daily returns on the All Ordinaries index shown in Figure 11.2. The time series is negatively skewed due to these two occurrences. Furthermore, there is a high degree of leptokurtosis evident which is common in most financial time series data.

Similar to the findings of Allen and Karjalainen (1994), skewness and kurtosis is greatly influenced by a small number of outliers. By removing those observations with an absolute value greater than 5 per cent greatly reduces skewness and kurtosis (see column 3 in Table 11.2). This is further reduced by removing those observations with an absolute value greater than 2.5 per cent (column 4 in Table 11.2).

Table 11.2: Descriptive statistics

	r_t (%)	$\lvert r_t \rvert < 0.05$ (%) (-9 obs.)	$\lvert r_t \rvert < 0.025$ (%) (-43 obs.)
Mean	0.0355	0.0524	0.0528
Maximum	5.6029	4.9801	2.4745
Minimum	-28.7590	-4.2244	-2.4980
Standard deviation	1.0181	0.8339	0.7817
Skew	-6.7657	-0.0507	-0.0681
Kurtosis	183.6861	4.9177	3.2730

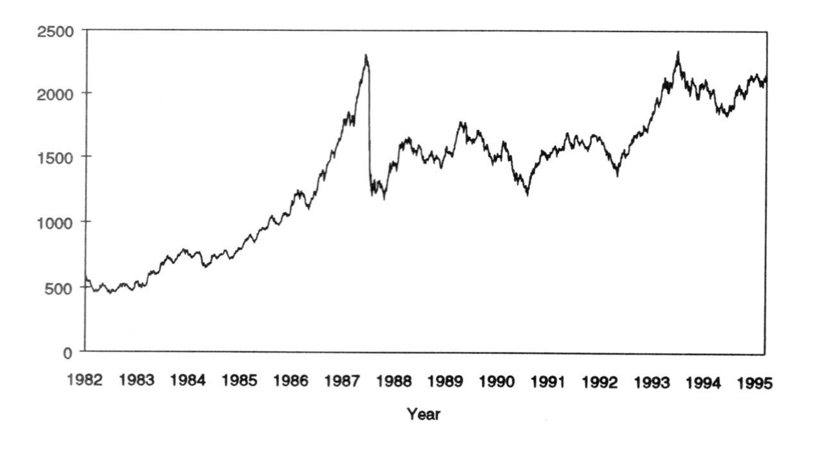

Figure 11.1: All Ordinaries index

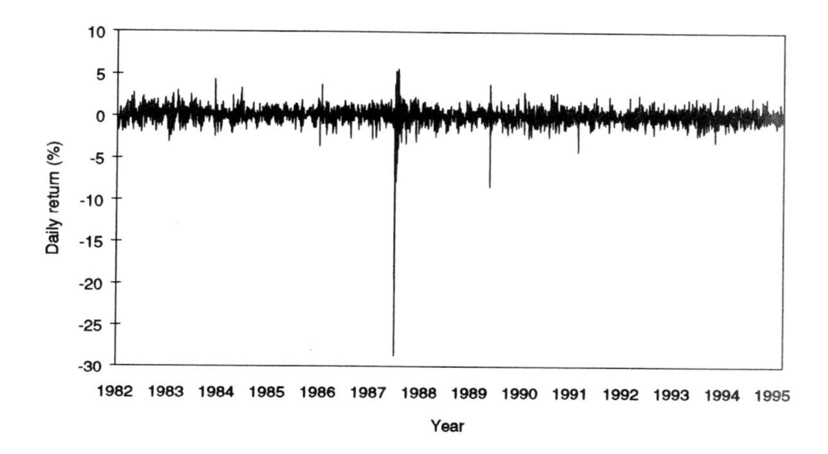

Figure 11.2: Continuously compounded returns

From the inspection of the autocorrelation function of daily returns on the All Ordinaries index shown in Table 11.3, it appears that there exists a significant degree of persistence in daily returns. This is indicated by significant positive autocorrelation found at various lags up to and including the tenth lag. These autocorrelations are too high to reconcile with the hypothesis of white noise. According to Taylor (1986), if the return series is white noise then the squared (R^2) and absolute (R^*) series should also be white noise. However, it is evident from Table 11.3 that this is not true since the autocorrelations in the absolute and squared returns appear to persist and do not die out as expected under a white noise process.

If the All Ordinaries index is assumed to follow a random walk, the persistence in the series of squared returns can be explained by standard autoregregressive conditional heteroscedasticity (ARCH) effects. The ARCH and GARCH models, first proposed by Engle (1982) and Bollerslev (1986) respectively, can be used to capture these effects.

11.3.3 Selection of Optimal Parameters

A genetic algorithm is programmed, using GAUSS, which is then used to search for the optimal technical trading rule parameter values using data

Table 11.3: Autocorrelations of daily returns

Lag	R	R^2	R^*
1	0.1194	0.0321	0.2290
2	-0.0268	0.0304	0.1789
3	0.0727	0.0734	0.2065
4	0.1082	0.0637	0.1795
5	0.0502	0.0657	0.1706
6	-0.0296	0.0411	0.1572
7	0.0685	0.0754	0.1866
8	0.0417	0.0050	0.0691
9	-0.0219	0.0337	0.1278
10	0.0524	0.0021	0.0607

$(\mathrm{CI} = \pm\frac{2}{\sqrt{N}} = \pm\ 0.0332)$

on the All Ordinaries index. Each of the rules requires a slightly different genetic algorithm due to differing problem representation and operation of the rules. For each trading rule one trial of the genetic algorithm is run, as summarised above in Table 11.1.[3]

Objective Function

The objective function is represented by trading rule profitability based on the total returns attributable to the use of a particular rule from trading in the All Ordinaries index. The rules used for trading are outlined in Section 11.1.

The rules developed for share market trading attempt to profit from successful market timing. The objective of a market timing strategy is to invest 100 per cent of funds available in the market with the highest expected return. In this study available funds are invested in either the Australian share market, in a diversified portfolio of shares weighted according to the All Ordinaries index, or in a risk-free security, 90-day RBA bills. The investment position taken depends on the current position. If the current position of the funds is in the share market a sell signal would result in the sale of shares

[3]Sensitivity analysis was performed on the parameter estimates by conducting ten trials of the *GA* for each rule. It was found that the estimates for the simple *MA*, channel and filter rules using the *GA* were optimal and did not vary across trials. The optimal values for these rules were found using an exhaustive search across the range of different parameter values. However, since an exhaustive grid-search procedure for the filtered and double *MA* would take many days it was not conducted. However it was found that the parameter estimates did not vary by much over ten trials of the GA.

and the purchase of bills with the proceeds from the sale, while a buy signal would lead to no action. If the current position of the funds is in bills a sell signal would result in no action, while a buy signal would lead to the sale of the bills and the purchase of shares. Technical trading rules based on a market timing strategy use technical analysis to decide whether to be in or out of the share market (market timing).

Rule profitability for trading in the share market is measured as excess returns, taking into account transaction costs explicitly and an adjustment for risk. Transaction costs (c) of 0.1 per cent are used since Allen and Karjalainen (1994) suggest that large institutional investors are able to achieve one-way transaction costs in the range of 0.1 to 0.2 per cent. Risk is considered by computing returns in excess of a buy and hold strategy. The excess return (R) given by:

$$R = R^{tr} - R^{bh} \tag{11.5}$$

is the return generated from using the trading rule (R^{tr}) above the return achieved from following a passive buy and hold strategy (R^{bh}).

A trading rule is a function that returns either a buy or sell signal. These signals divide the total number of trading days (N) into either *in* the market, earning the daily continuously compounded return (r_{mt}) or *out* of the market earning the daily return on the risk-free security (r_{ft}), as proxied by the 90-day RBA bill dealer rate. The return on the trading rule is:

$$R^{tr} = \sum_{t=2}^{N} d_t r_{mt} + \sum_{t=2}^{N} (1 - d_t)\, r_{ft} + \ln\left(\frac{1 - tc}{1 + tc}\right) T \tag{11.6}$$

which is composed of the summation of the daily market returns for days in the market and the daily returns on the risk-free security for days out of the market. Trading days are divided into days in or out of the market using a dummy variable d_t as defined in equation 11.3. An adjustment for transaction costs is given by the last term on the right-hand side of equation 11.6 which consists of the cost per transaction (tc) adjusted by the number of transactions (T). A position in the market on the final day of trading is forcibly closed out.

The return for the passive buy and hold strategy:

$$R^{bh} = \sum_{t=2}^{N} r_{mt} + \ln\left(\frac{1 - tc}{1 + tc}\right)$$

represents the total return on the market for the entire period of trading with an adjustment for a single transaction cost which is incurred on the last day of trading upon the sale of the market portfolio.

Table 11.4: Genetic algorithm parameters

Rule	b	p	c	m	$g^{\max}$
Simple MA	9	20	0.6	0.001	50
Filtered MA	19	50	0.6	0.001	60
Double MA	18	50	0.6	0.001	50
Filter	10	25	0.6	0.001	50
Channel	9	20	0.6	0.001	50

Genetic Algorithm Parameters

The genetic algorithm has five parameters $\{b, p, c, m, g^{\max}\}$, defined as:

b = number of elements in the vector
p = number of vectors or individuals in the population
c = probability associated with the occurrence of crossover
m = probability associated with the occurrence of mutation
$g^{\max}$ = maximum number of generations allowed

which can effect the functioning of the genetic algorithm. Small (large) values for b, p, m, g^{max} and a large (small) value of c result in rapid (slow) convergence; see Goldberg (1989) and Bauer (1994). The greater the rate of convergence, the lower the computational time that is required to reach the solution. However, if the rate of convergence is too rapid the solution space is not adequately searched, potentially missing the optimal solution (or better solutions to the one found).

Table 11.4 displays the parameters that are used by the GA for each of the trading rules. The choice of b values is discussed in subsection 11.2.1. However, the choice of p values was guided by previous studies (see Bauer, 1994) and experimentation with different values. The values for c and m are those recommended by Bauer (1994) based on sensitivity analysis conducted in previous studies.

11.3.4 Trading Rule Parameter Estimates

The results of the search for the optimal parameter values for the five different trading rules during the rule-building period using a genetic algorithm is given in Table 11.5. The annualised excess return on the share market during the rule-building period, given in column 2 of Table 11.5, is very high for all five trading rules. However, these extremely high returns should be viewed with a degree of scepticism as this seven-year period includes a period of very

Table 11.5: Trading rule parameter estimates

Rule	Return (%)	Parameter estimates	Time	M[a]	G[b]
Simple MA	16.96	$L = 10$	12 m & 55 s	5	37
Filtered MA	16.79	$L = 12$ & $X = 0.01$	42 m & 15 s	60	60*
Double MA	16.96	$S = 1$ & $L = 10$	41 m & 3 s	45	50*
Filter	13.13	$X = 0.80$	18 m & 35 s	9	45
Channel	15.98	$L = 9$	17 m & 49 s	9	46

(a) M refers to the number of mutations.

(b) G refers to the number of generations simulated.

high returns during the mid-1980s prior to the share market crash of 1987 and the mini-crash of 1989. Furthermore, the trading rules might have been able to profit from predicting these two significant events.

The genetic algorithm converges quickly for the simple trading rules (simple MA, filter and channel rules), taking on average approximately 15 minutes using a 486 DX2 66 computer. An exhaustive search procedure takes on average approximately 20 minutes for the simple MA and channel rules which have only 500 different possible parameter values, and 40 minutes for the filter rule which has 1000 different possible values. For these three rules it is obvious that the genetic algorithm offers little in terms of speed. This property of efficiency in computing time is better appreciated in problems which have very large solution spaces where an exhaustive search procedure would take days if not weeks to find the optimum. The two more complex rules (filtered MA and the double MA) which have a much larger solution space, 500,500 and 250,000 possible parameters values respectively, provide a greater justification for the use of a genetic algorithm. On average the genetic algorithm took approximately one hour, whereas an exhaustive grid-search procedure would take many days.

Other mathematical optimisation techniques such as gradient-type and scoring algorithms can be employed. However, in problems with multiple optima and extremely large solution spaces these traditional methods might not be as successful as a genetic algorithm in getting close to the optimum, since such methods are prone to becoming stuck at local optima (see Dorsey and Mayer, 1995).

11.3.5 Out-of-sample Performance

The performance of the trading rules out-of-sample is given in Table 11.6. The profitability of the trading rules out-of-sample is much lower than with the rule-building sample. Share market trading rules yielded an average out-of-sample excess return (R) of 2.9 per cent annually compared to 16 per cent in sample. The MA and channel rules appear to have outperformed the filter rule. The rules resulted in 107 transactions (T) on average.

The average of the cumulative returns from all five trading rules on the share market, shown in Figure 11.3, is nearly always positive for the trading rules and above the buy-and-hold strategy. Furthermore the cumulative return of the buy-and-hold strategy is negative for the first half of the period.

These results indicate that small positive excess returns are achievable using share market trading rules employing the parameter estimates from the genetic algorithm. However, these excess returns appear to be insignificant due to the results of the tests of significance. The t ratio (t_b), given in column 5 of Table 11.6, tests whether the difference between the average return attributable to days in the market ($\bar{r}_b$) and the unconditional mean return on the market is significantly different from zero. The t ratio (t_s), given in column 7 of Table 11.6, tests whether there exists a significant difference between the average return attributable to days out of the market ($\bar{r}_s$) and the unconditional mean return on the market. The first test seeks to examine whether the trading rule leads to being in the market on the days when the market performs well, while the second test examines whether the trading rule leads to being out of the market on the days when the market performs poorly. Most of these tests show that the results are not statistically significant at either the 5 per cent or 10 per cent level of significance. The last column of Table 11.6 reports the t ratios from the test of significance of the difference between the average of returns received from being in the market and the average of returns obtained from being out of the market. The results for this test, although not overwhelming, are significant for all the rules except the channel rule. This result supports the hypothesis that returns attributable to days in the market are higher than those when out of the market, implying that the rules have some market-timing ability.

Also evident from the standard deviations of the daily returns on the trading rules, which appear in brackets in the fourth and sixth columns in Table 11.6, is that the variability of returns to days in the market are lower than when out of the market. This would tend to suggest that the trading strategies based on the share market rules investigated are less risky than a passive buy-and-hold strategy.

However, these tests of significance, which under the null hypothesis are based on the assumption of normality, might not be appropriate due to the

Table 11.6: Trading rule excess returns

Rule	R	T	$\overline{r}_b$	t_b	$\overline{r}_s$	t_s	t
Simple MA	2.06	124	0.0577	1.1716	-0.0245	-1.2023	2.0559
			(0.7514)		(0.8182)		
Filtered MA	4.77	102	0.0728	1.6047	-0.0393	-1.6299	2.8013
			(0.7379)		(0.8284)		
Double MA	2.06	124	0.0577	1.1716	-0.0245	-1.2023	2.0559
			(0.7514)		(0.8182)		
Filter	1.81	134	0.0573	1.1735	-0.0271	-1.2618	2.1090
			(0.7723)		(0.7985)		
Channel	3.54	50	0.0507	0.9759	-0.0169	-0.9751	1.6903
			(0.7589)		(0.8106)		
Average	2.85	107	0.0592		-0.0265		
			(0.7598)		(0.8148)		

non-normality of daily returns for both the share market and the AUD/USD exchange rate as discussed in subsection 11.3.2. Brock et al. (1992) suggest using the bootstrapping technique to test the statistical significance of the excess returns, which does not depend upon distributional assumptions.

11.4 Conclusions

The share market technical trading rules employing the genetic algorithm derived parameter estimates generate small positive excess returns out-of-sample compared to in-sample. These rules, based on a market timing strategy, outperform a passive buy-and-hold strategy. Furthermore, the average cumulative daily returns on the rules is nearly always positive. From a trading perspective these are encouraging results.

There are several possible reasons why trading rule profits are lower out-of-sample than compared to in-sample. First, it could be that financial markets have become more efficient over time, as reflected by the diminished profits in the out-of-sample compared to the in-sample period. Another possible reason is that the trading rules may have been able to profit from the stock market crashes of 1987 and 1989. Alternatively, it may be the case that there is limited usefulness in optimising trading rule parameters using historical data. Lukac and Brorsen (1989) come to such a conclusion in a study of the futures markets using a channel rule and directional movement system.

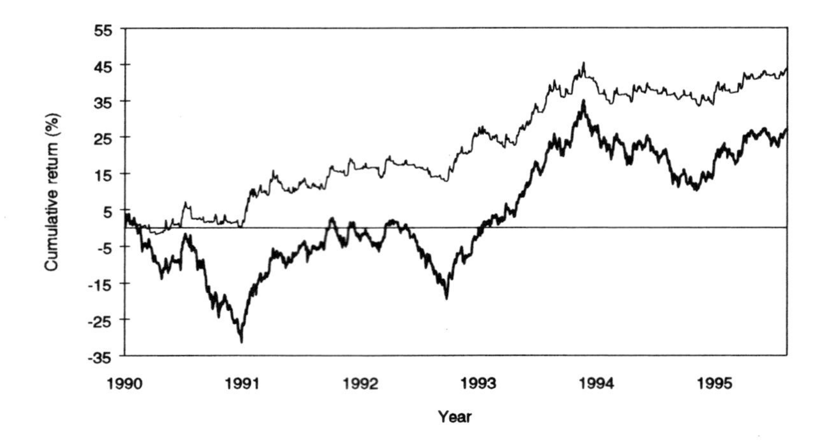

Figure 11.3: Average return (not bold) and buy-hold return (bold)

The profits generated from the trading rules appear to be statistically insignificant according to standard t tests. However, since these tests are based on the assumption of normality, which is clearly not valid for the share market data investigated in this study, no conclusion concerning the significance of the rules is made at this stage. Future research will investigate this issue using the more appropriate bootstrap methodology suggested by Brock et al. (1992) which does not restrict the distribution of the returns under the null hypothesis.

Part IV

Neural Network Applications

Chapter 12

Artificial Neural Networks

Vance L. Martin and Clarence Tan

Artificial neural networks are flexible functional forms which are used to model a broad range of processes within a large number of disciplines. The term 'neural network' acts as a reminder that this class of models is motivated by the way the brain processes information, whereas the use of the word 'artificial' recognises that these models represent at best an approximation of brain behaviour. For example, a brain has about one hundred billion neurons, whereas artificial neural networks would have at most a few hundred neurons and in most econometric applications the number would be much less than one hundred. The comparable situation in econometrics is that the brain is the economy consisting of about one hundred billion microeconomic structural equations, where each structural equation acts as a neuron which combines information by weighting variables with the magnitude of the weights determined by estimated parameters. However, as with artificial neural networks, in practice at most a small proportion of structural microeconomic equations, or a set of macroeconomic equations, are used to model the overall economy. This would suggest that in neural network terminology, econometric modelling is 'artificial'.

Research in artificial neural networks dates from the work of McCulloch and Pitts (1943), but the development of high speed computers during the last decade has enabled artificial neural networks to gain greater popularity. A first glance into the neural network literature leaves two impressions. First, there are extravagant claims made about the forecasting power of artificial neural networks in a broad range of disciplines, ranging from economics and finance to psychology and artificial intelligence; for a survey of neural network applications in finance, see Trippi and Turbin (1996). Secondly, there is a vast amount of jargon which has the tendency to make the discipline as mysterious and just as difficult to understand as the brain. Phrases such as 'training the network', 'learning and intelligence', 'activation and squasher functions', 'connection strengths', 'hidden layers' and 'back propagation', to name just a few, all add to the mystique.

From an econometric perspective artificial neural networks represent a flexible framework for building and estimating nonlinear time series and structural models. Artificial neural network parlance does not contain the terms modelling and estimation, but there are parallels which become clear when neural networks are formalised and the computer algorithms used to 'train the network' are explicitly stated. This chapter is designed to provide a guide to artificial neural networks by translating the features of neural network structures and modelling strategies into standard econometric methodology; namely, model specification, estimation and testing. In drawing parallels between the two fields this serves to highlight that much, if not all, standard econometric practice can be given a neural network interpretation; see also White (1992), Kuan and White (1994) and Sarle (1994). This chapter also provides a theoretical framework for the empirical studies using neural networks in chapters 13 and 14.

Section 12.1 states the model specification features of artificial neural networks and shows them to be related to both linear and nonlinear econometric models. Estimation issues are discussed in Section 12.2 with special emphasis on showing the relationship between the workhorse of neural network computer models, the back propagation algorithm, and nonlinear estimation methods used in econometrics. Testing issues are discussed in Section 12.3 with special emphasis on constructing Lagrange multiplier tests of artificial neural networks based on the framework of Lee *et al.* (1993). Relationships between artificial neural networks and some specific econometric models are discussed in Section 12.4. In Section 12.5 an artificial neural network trading model for the foreign exchange market is constructed.

12.1 Model Specification

12.1.1 Basics and Terminology

The salient features of neural network models can be highlighted within the context of nonlinear time series models. To begin with, consider the linear AR(1) time series model:

$$Y_t = \alpha_0 + \alpha_1 Y_{t-1} + v_t \tag{12.1}$$

where Y_t is a stochastic process, α_i are parameters and v_t is white noise. The parameters α_0 and α_1 can be estimated by OLS by regressing Y_t on $\{1, Y_{t-1}\}$.

The main characteristic of neural network models is that the relationship between Y_t and Y_{t-1} is allowed to be nonlinear. A typical neural network specification consists of augmenting (12.1) by a logistic function:

$$Y_t = \alpha_0 + \alpha_1 Y_{t-1} + \phi_1 F_t + u_t \tag{12.2}$$

where u_t is an error term and F_t is given by:

$$F_t = \frac{1}{1 + \exp\left[-\left(\beta_0 + \beta_1 Y_{t-1}\right)\right]} \tag{12.3}$$

and the parameter set is $\{\alpha_0, \alpha_1, \phi, \beta_0, \beta_1\}$. As discussed at length below, the neural network parameters can be estimated by choosing the parameters β_0 and β_1 randomly, thereby creating a time series on F_t, and estimating the remaining parameters $\alpha_0, \alpha_1, \phi_1$, by regressing Y_t on $\{1, Y_{t-1}, F_t\}$.

In the neural network literature, the nonlinear time series model given by (12.2) and (12.3) represents a hidden layer feedforward artificial neural network. The hidden layer refers to the logistic function F_t in (12.3). This function is also referred to as a squasher or activation function. In neural network parlance Y_{t-1} is referred to as the input and Y_t as the target variable. In time series models these are respectively the independent and dependent variables. The predictor from (12.2) and (12.3):

$$Y_t^* = \alpha_0 + \alpha_1 Y_{t-1} + \phi_1 F_t \tag{12.4}$$

is called the output in neural network models. Finally, the intercepts $\{\alpha_0, \beta_0\}$ are often called the bias.

The artificial neural network considered in (12.2) and (12.3) can be simplified in several ways as the discussion that follows demonstrates. In econometrics, the choice of functional forms, variables and lag lengths all come under the banner of model specification. In neural network parlance this is referred to as architecture.

12.1.2 Univariate Generalisations

The univariate nonlinear time series model discussed so far can be generalised in several ways:

1. additional lags to allow for richer dynamic structures,
2. additional logistic functions to allow for more hidden layers,
3. the inclusion of multilayers,
4. the adoption of alternative nonlinear functions.

The first two features are accommodated by extending (12.2) and (12.3) as:

$$Y_t = \alpha_0 + \sum_{i=1}^{K} \alpha_i Y_{t-i} + \sum_{i=1}^{N} \phi_i F_{i,t} + u_t \tag{12.5}$$

where:

$$F_{i,t} = \frac{1}{1 + \exp\left[-\left(\beta_{i,0} + \sum_{j=1}^{M} \beta_{i,j} Y_{t-j}\right)\right]} \tag{12.6}$$

The number of hidden layers is N and the number of lags are determined by K and M. This model is a nonlinear AR(K).

The lag structures in (12.5) and (12.6) are finite. To allow for longer lag structures which allow for parsimonious parameterisations, $F_{i,t}$ is expressed as a function of $F_{j,t-k}$, $\forall\ j, k > 0$. For example, (12.6) is extended as:

$$F_{i,t} = \frac{1}{1 + \exp\left[-\left(\beta_{i,0} + \sum_{j=1}^{M} \beta_{i,j} Y_{t-j} + \sum_{j=1}^{N} \sum_{k=1}^{M} \delta_{i,j,k} F_{j,t-k}\right)\right]} \tag{12.7}$$

The inclusion of the $F_{j,t-k}$ terms in (12.7) is comparable to the inclusion of moving average terms in AR models. This class of neural network models is referred to as recurrent networks, whereas the feedforward class of networks discussed so far is given by setting $\delta_{i,j,k} = 0, \forall\ i, j, k > 0$.

To allow for multilayered hidden functions, it is necessary to express $F_{i,t}$ in terms of other logistic functions. For example, the logistic function in (12.6) can be re-expressed as:

$$F_{i,t} = \frac{1}{1 + \exp\left[-\left(\beta_{i,0} + \sum_{j=1}^{M} \beta_{i,j} G_{j,t-j}\right)\right]} \tag{12.8}$$

where $G(.)$ is another logistic function given by:

$$G_{j,t} = \frac{1}{1 + \exp\left[-\left(\gamma_{j,0} + \sum_{k=1}^{L} \gamma_{j,k} Y_{t-j}\right)\right]} \tag{12.9}$$

The nonlinearities discussed so far are in the form of a logistic function. However, a range of possibilities are available. Some examples are as follows:

1. Linear:

$$H(s) = s \tag{12.10}$$

2. Logistic

$$H(s) = \frac{1}{1 + \exp[-s]} \tag{12.11}$$

3. Hyperbolic tangent

$$H(s) = \tanh(s) \tag{12.12}$$

4. Threshold

$$H(s) = \begin{cases} 1: & s > \xi \\ 0: & s \leq \xi \end{cases} \tag{12.13}$$

where ξ is a constant, say 0.0 or 0.5.

5. Gaussian

$$H(s) = \frac{1}{\sqrt{2\pi}} \exp\left[-0.5s^2\right] \tag{12.14}$$

The logistic equation represents the most commonly used function for introducing nonlinearities and hence for modelling hidden layers. The hyperbolic tangent is functionally related to the logistic function and thus is also well used in modelling hidden layers. The threshold function, also known as the heaveside function, represents the earliest function for modelling hidden layers. The use of the Gaussian function gives rise to radial basis neural networks. This class of models is also related to the nonparameteric kernel regression estimator of Nadaraya (1964) and Watson (1964).

The linear function given by (12.10) serves as a way of nesting linear models. It also implies that the nonlinear time series model in (12.2) can be interpreted as having two hidden layers, one linear and one logistic. Further generalisations can be entertained by specifying (12.2) as:

$$Y_t = H\left(\alpha_0 + \alpha_1 Y_{t-1} + \phi_1 F_t\right) + u_t \tag{12.15}$$

where $H(.)$ is one of the functions listed in (12.10) to (12.14). For the linear case given by (12.10) where $H(s) = s$, (12.15) reduces to (12.2).

12.1.3 Multivariate Generalisations

The class of models discussed so far are formulated for a single variable Y_t. Further generalisations are:

1. the inclusion of several exogenous variables $X_{i,t}$,

2. the inclusion of several endogenous variables $Y_{i,t}$.

The addition of exogenous variables can give the neural network models a structural interpretation. For example, consider the single equation structural model relating $Y_{1,t}$ and $X_{1,t}$:

$$Y_{1,t} = \alpha_0 + \alpha_1 X_{1,t} + w_t \tag{12.16}$$

where w_t is an error term. Now following the neural model specifications above, a nonlinear augmentation of this structural model is:

$$Y_{1,t} = \alpha_0 + \alpha_1 X_{1,t} + \phi_1 F_t + u_t \tag{12.17}$$

where u_t is an error term and F_t is given by:

$$F_t = \frac{1}{1 + \exp\left[-\left(\beta_0 + \beta_1 X_{1,t}\right)\right]} \tag{12.18}$$

The exogenous variable $X_{1,t}$ represents the input in neural network models. By allowing for a dynamic structural model through the inclusion of lags of $Y_{1,t}$ and $X_{1,t}$, the set of inputs is given by $\{Y_{1,t-1}, Y_{1,t-2}, ...; X_{1,t}, X_{1,t-1}, ...\}$. By extending the number of exogenous variables to $X_{1,t}, X_{2,t}, ...$, in (12.17) and (12.18), a multivariate nonlinear single equation structural model can be formulated which generalises the linear regression model.

Multivariate neural networks can be formulated which can be interpreted as multivariate nonlinear dynamic structural models which nest multivariate linear structural models. As an example, consider the following nonlinear bivariate model in $Y_{1,t}$ and $Y_{2,t}$:

$$\begin{aligned} Y_{1,t} &= \alpha_{1,0} + \alpha_{1,1} Y_{1,t-1} + \alpha_{1,2} Y_{2,t-1} + \phi_1 F_{1,t} + u_{1,t} \\ Y_{2,t} &= \alpha_{2,0} + \alpha_{2,1} Y_{1,t-1} + \alpha_{2,2} Y_{2,t-1} + \phi_2 F_{2,t} + u_{2,t} \end{aligned} \tag{12.19}$$

where $u_{1,t}$ and $u_{2,t}$ are error terms, and $F_{1,t}$ and $F_{2,t}$ are logistic functions given by respectively:

$$\begin{aligned} F_{1,t} &= \frac{1}{1 + \exp\left[-\left(\beta_{1,0} + \beta_{1,1} Y_{1,t-1} + \beta_{1,2} Y_{2,t-1}\right)\right]} \\ F_{2,t} &= \frac{1}{1 + \exp\left[-\left(\beta_{2,0} + \beta_{2,1} Y_{1,t-1} + \beta_{2,2} Y_{2,t-1}\right)\right]} \end{aligned} \tag{12.20}$$

Setting $\phi_1 = \phi_2 = 0$ in (12.19) yields as a special case the reduced form of a linear bivariate structural model. This special case is also known as a bivariate VAR.

A further extension of (12.19) is to allow for a recursive structure between $Y_{1,t}$ and $Y_{2,t}$. For example, rewriting (12.19) as:

$$\begin{aligned} Y_{1,t} &= \alpha_{1,0} + \alpha_{1,1} Y_{1,t-1} + \alpha_{1,2} Y_{2,t-1} + \phi_1 F_{1,t} + \delta Y_{2,t} + u_{1,t} \\ Y_{2,t} &= \alpha_{2,0} + \alpha_{2,1} Y_{1,t-1} + \alpha_{2,2} Y_{2,t-1} + \phi_2 F_{2,t} + u_{2,t} \end{aligned} \tag{12.21}$$

allows for a contemporaneous relationship from $Y_{2,t}$ to $Y_{1,t}$.

12.2 Estimation

This section discusses three algorithms for estimating the parameters of neural networks. These are: search methods, gradient methods and the back propagation method. As neural network models are nonlinear it is necessary to use an iterative algorithm. It is shown below that it is this feature of the problem of estimation that in neural network circles is commonly referred to as learning.

For a sample of $t = 1, 2, ..., T$, observations on the variable Y_t, the aim of the estimation algorithms is to find a set of parameters Ψ which solves:

$$\min_{\Psi} \frac{1}{2} \sum_{t=1}^{T} (Y_t - Y_t^*)^2 \tag{12.22}$$

where Y_t^* is the predictor and the constant $1/2$ is chosen to simplify the derivations below. This is just the standard least squares problem of minimising the sum of squared errors. In neural network parlance, this expression is commonly referred to as either the fitness function, or the cost function, or even the penalty function. For certain applications of artificial neural networks alternative objective functions are used. For example, in the development of a foreign exchange trading system investigated in Section 12.5, the objective function is to maximise the profits from trading which, in turn, includes a range of variables such as transaction costs.

12.2.1 Search Methods

The simplest algorithm is to choose the parameters inside the activation function randomly and estimate the remaining parameters by OLS. For example, to estimate the parameters $\Psi = \{\alpha_0, \alpha_1, \phi, \beta_0, \beta_1\}$ in (12.2) and (12.3), the approach is to choose the parameters $\{\beta_0, \beta_1\}$ from a random number generator and create the time series F_t according to (12.3). The remaining parameters $\{\alpha_0, \alpha_1, \phi_1\}$ in (12.2) are obtained by regressing Y_t on $\{1, Y_{t-1}, F_t\}$. By repeating the search algorithm several times, the set of parameters chosen is the set that minimises the objective function in (12.22). The big advantage of this algorithm is that it just requires a random number generator and a least squares routine, which are available in all statistical packages.

To investigate the success of the search algorithm, Monte Carlo experiments are conducted using the following four models:

$$\begin{array}{ll} \text{Deterministic chaos:} & Y_t = 3.8Y_{t-1}\left(1 - Y_{t-1}\right) \\ \text{Bier-Bountis map:} & Y_t = -2 + 28.5\frac{Y_{t-1}}{1+Y_{t-1}^2} \\ \text{Bilinear model:} & Y_t = 0.7Y_{t-1}v_{t-1} + v_t \\ \text{Stochastic chaos:} & Y_t = w_t 4Y_{t-1}\left(1 - Y_{t-1}\right) \end{array} \tag{12.23}$$

where $v_t \sim N(0,1)$ and $w_t \sim U(0,1)$. The sample size is $T = 250$. All models are simulated for $T + 100$ periods with the first 100 observations deleted to overcome start-up problems. The first two models are deterministic which are potentially perfectly predictable. These two models are also investigated by Kuan and White (1994) thereby providing a source of comparison with their results. The next two models are stochastic and thus are not perfectly predictable. The stochastic chaos model is equivalent to the standard logistic map except that now the parameter is allowed to vary from 0 to 4, at each t. This means that Y_t can exhibit chaotic behaviour for certain time periods, whilst switching to cyclical, or even monotonic behaviour for other time periods.

The neural network model estimated in all experiments is a hidden layer feedforward artificial neural network:

$$Y_t = \alpha_0 + \sum_{i=1}^{L} \alpha_i Y_{t-i} + \sum_{i=1}^{N} \phi_i \left(\frac{1}{1 + \exp\left[-\left(\beta_{i,0} + \sum_{j=1}^{L} \beta_{i,j} Y_{t-j}\right)\right]} \right) + u_t \tag{12.24}$$

where u_t is an error term, the lag lengths are chosen as $L = 1, 2$, and the number of activation functions range from $N = 0, 1, ..., 10$. The zero activation case, $N = 0$, represents the linear model and thus serves as a benchmark for comparing the ability of the neural network to capture the nonlinearities in the data over and above the linear model. The random number generator used in the search algorithm to compute the $\beta_{i,j}$ values is uniform over the range $[-2, 2]$.

Tables 12.1 to 12.4 give the results of estimating the artificial neural network using the search algorithm based on 100 random searches. The statistics reported are: the sum of squared errors (SSR), the coefficient of determination (R^2), and the Schwartz information criterion ($\text{SIC} = \ln \widehat{\sigma} + K \ln(T)/2T$, where K is the number of estimated parameters, $\widehat{\sigma}$ is the standard error of estimate and $T = 250$). The SIC has the advantage over the R^2 as the addition of parameters in the network are penalised thereby helping to overcome

Table 12.1: Deterministic chaos

N	$L=1$			$L=2$		
	SSR	R^2	SIC	SSR	R^2	SIC
0	8.699	0.442	-1.655	8.045	0.481	-1.681
1	0.000	1.000	-12.847	0.033	0.998	-4.390
2	0.000	1.000	-12.581	0.001	1.000	-7.176
3	0.000	1.000	-13.501	0.000	1.000	-8.218
4	0.000	1.000	-13.577	0.000	1.000	-10.632
5	0.000	1.000	-11.797	0.000	1.000	-11.530
6	0.000	1.000	-12.736	0.000	1.000	-11.615
7	0.000	1.000	-7.804	0.000	1.000	-11.546
8	0.000	1.000	-9.038	0.014	0.999	-10.874
9	0.302	0.981	-3.035	0.001	1.000	-8.436
10	0.159	0.990	-3.322	0.003	1.000	-11.435

the curse of dimensionality problem. The results in Table 12.1 for the deterministic chaos model show that the search algorithm works very well as the neural network generates a perfect fit with $N = 1$ and $L = 1$.[1] For the Bier-Bountis model in Table 12.2 a near perfect fit is achieved for $N = 3$ and $L = 1$. Increasing N to $N = 8$ results in the sum of squared residuals equaling $SSR = 0.0$. These results also highlight an important feature of neural network models, namely that for deterministic processes this class of models can yield a perfect fit.

The results for the two stochastic models in Tables 12.3 and 12.4 show that the neural network significantly improves upon the linear model, $N = 0$. Using the SIC criteria, the best neural network model is $N = 1$ and $L = 2$ for the bilinear model, and $N = 1$ and $L = 1$ for the stochastic chaos model.

12.2.2 Gradient Methods

The combination of a nonlinear model and a sum of squares objective function suggests the use of standard gradient algorithms such as Newton-Raphson, or more specifically, a nonlinear least squares procedure. To highlight these algorithms, consider the following neural network model which relates the

[1] The expression a 'perfect fit' is relative to rounding error and machine accuracy. Also note that some of the results are not always consistent for certain parameterisations. For example, increasing the number of activation functions from $N = 8$ to $N = 9$ in Table 12.1 with $L = 1$, results in a small reduction in R^2. This reflects that the search algorithm is restricted to 100 searches and thus the error sum of squares has not been minimised.

Table 12.2: Bier-Bountis map

N	$L=1$			$L=2$		
	SSR	R^2	SIC	SSR	R^2	SIC
0	739.643	0.859	0.566	730.589	0.860	0.574
1	512.165	0.902	0.416	54.951	0.989	-0.676
2	4.742	0.999	-1.892	37.356	0.993	-0.824
3	0.138	1.000	-3.626	5.774	0.999	-1.713
4	0.032	1.000	-4.330	4.143	0.999	-1.835
5	0.015	1.000	-4.655	0.674	1.000	-2.698
6	0.002	1.000	-5.605	0.126	1.000	-3.492
7	0.001	1.000	-5.967	0.054	1.000	-3.871
8	0.000	1.000	-6.944	0.044	1.000	-3.933
9	0.000	1.000	-7.646	0.037	1.000	-3.968
10	0.000	1.000	-7.944	0.006	1.000	-4.805

Table 12.3: Bilinear model

N	$L=1$			$L=2$		
	SSR	R^2	SIC	SSR	R^2	SIC
0	715.634	0.045	0.550	706.697	0.055	0.557
1	633.247	0.155	0.522	596.317	0.203	0.516
2	605.578	0.192	0.533	550.009	0.264	0.520
3	597.062	0.203	0.559	566.109	0.243	0.579
4	582.005	0.223	0.580	508.407	0.320	0.570
5	553.133	0.262	0.587	508.347	0.320	0.614
6	540.467	0.279	0.609	493.273	0.340	0.644
7	535.771	0.285	0.638	460.451	0.384	0.654
8	531.816	0.290	0.667	455.001	0.391	0.692
9	528.651	0.294	0.698	425.847	0.430	0.704
10	527.511	0.296	0.730	427.707	0.428	0.750

Table 12.4: Stochastic chaos

N	$L=1$			$L=2$		
	SSR	R^2	SIC	SSR	R^2	SIC
0	11.645	0.115	-1.509	11.566	0.121	-1.499
1	8.618	0.345	-1.626	8.602	0.346	-1.603
2	8.600	0.347	-1.594	8.541	0.351	-1.562
3	8.539	0.351	-1.564	8.555	0.350	-1.517
4	8.530	0.352	-1.532	8.541	0.351	-1.473
5	8.525	0.352	-1.499	8.515	0.353	-1.430
6	8.526	0.352	-1.466	8.505	0.354	-1.386
7	8.535	0.352	-1.432	8.490	0.355	-1.343
8	8.534	0.352	-1.399	8.438	0.359	-1.301
9	8.542	0.351	-1.365	8.376	0.364	-1.260
10	8.707	0.339	-1.322	8.390	0.362	-1.215

dependent variable Y_t and the two independent variables $X_{1,t}$ and $X_{2,t}$:

$$Y_t = \frac{1}{1+\exp\left[-\left(\beta_1 X_{1,t} + \beta_2 X_{2,t}\right)\right]} + u_t \tag{12.25}$$

where $\Psi = [\beta_1, \beta_2]'$ is a (2×1) vector of unknown parameters, and u_t is an error term. Given that the objective function is to minimise the sum of squared errors in (12.22), where the predictor, or output in neural network terminology, is:

$$Y_t^* = \frac{1}{1+\exp\left[-\left(\beta_1 X_{1,t} + \beta_2 X_{2,t}\right)\right]} \tag{12.26}$$

a potential iterative algorithm is to update the parameter vector at the $(k+1)^{\text{th}}$ iteration according to:

$$\begin{aligned} \Psi^{(k+1)} &= \Psi^{(k)} - \left.\frac{\partial C}{\partial \Psi}\right|_{\Psi=\Psi^{(k)}} && (12.27) \\ &= \Psi^{(k)} + \sum_{t=1}^{T} \begin{bmatrix} \left(Y_t - Y_t^*\right) f_t X_{1,t} \\ \left(Y_t - Y_t^*\right) f_t X_{2,t} \end{bmatrix}_{\Psi=\Psi^{(k)}} && (12.28) \end{aligned}$$

where f_t is the logistic function which is evaluated at the kth iteration:

$$f_t = \frac{\exp\left[-\left(\beta_1^{(k)} X_{1,t} + \beta_2^{(k)} X_{2,t}\right)\right]}{\left(1+\exp\left[-\left(\beta_1^{(k)} X_{1,t} + \beta_2^{(k)} X_{2,t}\right)\right]\right)^2} \tag{12.29}$$

Convergence is achieved when $\Psi^{(k+1)} \simeq \Psi^{(k)}$, which is equivalent to $\frac{\partial C}{\partial \Phi}\big|_{\Psi=\Psi^{(k+1)}} \simeq 0$. For certain iterations $C\left(\Psi^{(k+1)}\right) > C\left(\Psi^{(k)}\right)$, so that the updated parameters do not reduce the sum of squared errors, but increase it. To help guard against this, a line search parameter, η, is commonly introduced as follows:

$$\Psi^{(k+1)} = \Psi^{(k)} - \eta \left.\frac{\partial C}{\partial \Phi}\right|_{\Psi=\Psi^{(k)}} \tag{12.30}$$

For well-behaved problems, $\eta = 1$ and the algorithm proceeds as before. For non-well-behaved problems, η is chosen to ensure that there is an improvement in the objective function. A simple approach is to reduce η, say, from 1.0 to 0.5, to 0.25 and so on, until an improvement is achieved. This process is sometimes referred to as squeezing and η is referred to as the squeezing parameter.[2]

The algorithm in (12.30) is known as the method of steepest descent which just makes use of the vector of first order derivatives of the objective function (12.22). Compared to other algorithms which use second order derivatives, the method of steepest descent converges more slowly. To allow for second derivatives the iteration step is:

$$\Psi^{(k+1)} = \Psi^{(k)} - \eta \left. H^{-1}G\right|_{\Psi=\Psi^{(k)}} \tag{12.31}$$

where $G = \partial C/\partial \Psi$ is the gradient vector of the objective function (12.22), evaluated at $\Psi^{(k)}$, and H is a matrix that is determined by the choice of algorithm. For the model in (12.25):

$$G = -\sum_{t=1}^{T} \left[\begin{array}{c} (Y_t - Y_t^*)\, f_t X_{1,t} \\ (Y_t - Y_t^*)\, f_t X_{2,t} \end{array} \right] \tag{12.32}$$

where f_t is given by (12.29). In the method of steepest descent $H = I$, whereas in the Newton-Raphson method H is the matrix of second derivatives of (12.22).[3] To estimate neural network parameters by nonlinear least squares, $H = \sum_t Z_t' Z_t$ where $Z_t = [f_t X_{1,t},\ f_t X_{2,t}]$. Combining with (12.32), $H^{-1}G|_{\Psi=\Psi^{(k)}}$ is computed by regressing the error $Y_t - Y_t^*$ on $\{f_t X_{1,t},\ f_t X_{2,t}\}$ where all parameters are evaluated at $\Psi^{(k)}$. Alternatively, H is chosen by using the Broyden, Fletcher, Goldfarb and Shanno (BFGS) algorithm which

[2] For a discussion of safeguards to achieve the global minimum, see Harvey (1990). One approach, similar to the search algorithm discussed above, is to choose the parameters randomly when the updated parameter vector does not lower the sum of squared errors in (12.22).

[3] In the method of steepest descent, the line search parameter η in (12.31) is a function of the matrix of second derivatives as this ensures that the direction is steepest.

provides a rank one updating formula for computing H^{-1}; see also Kuan and White (1994, pp.46-48).

The iterative algorithms discussed here are all available in standard econometric packages such as EVIEWS, RATS, SHAZAM and TSP, or in more sophisticated programs such as GAUSS and MATLAB, and thus can be used to estimate the parameters of artificial neural networks.

12.2.3 Back Propagation

The back propagation algorithm is perhaps the most commonly used method for estimating neural network parameters. The basic back propagation algorithm is similar to the method of steepest descent as only information on first derivatives is used, with the exception that the cost function is just evaluated at time $t = k$, and not the sum over all t:

$$\Psi^{(k+1)} = \Psi^{(k)} - \left.\frac{\partial C_t}{\partial \Phi}\right|_{\Psi=\Psi^{(k)}} \tag{12.33}$$

where $C_t = 0.5\left(Y_t - Y_t^*\right)^2$. For example, to estimate (12.25) by the method of back propagation, the updating rule given by (12.27) and hence (12.28) becomes:

$$\Psi^{(k+1)} = \Psi^{(k)} + \left[\begin{array}{c} \left(Y_k - Y_k^*\right) f_k X_{1,k} \\ \left(Y_k - Y_k^*\right) f_k X_{2,k} \end{array}\right]_{\Psi=\Psi^{(k)}} \tag{12.34}$$

The algorithm begins by specifying the initial parameters $\Psi^{(0)}$, say randomly. For a data set defined over $t = 1, 2, ..., T$ observations, there are T iterations. Convergence occurs as with the gradient algorithms, where $\Psi^{(k+1)} \simeq \Psi^{(k)}$. If convergence is not achieved after T iterations, the process is repeated with the initial parameter vector $\Psi^{(0)}$ now being based on the most recent parameter values, namely $\Psi^{(T)}$. In neural network terminology, repeating the process through the data set from $t = 1, 2, ..., T$ is known as a training epoch. If the maximum number of training epochs is 100, then the maximum number of iterations is $100T$.

Two extensions of the back propagation algorithm given in (12.33) are given by:

$$\Psi^{(k+1)} = \Psi^{(k)} - \eta \left.\frac{\partial C_t}{\partial \Phi}\right|_{\Psi=\Psi^{(k)}} + \zeta\left(\Psi^{(k)} - \Psi^{(k-1)}\right) \tag{12.35}$$

The parameters η and ζ are commonly referred to respectively as the 'learning rate' and the 'momentum or bump coefficient' in neural network terminology. Comparing (12.35) with (12.30) shows that the learning rate is just simply the line search, or squeezing parameter. White (1989) and Kuan and White (1994) show that by choosing the learning parameter as:

$$\eta \propto t^{-n} \qquad 0.5 < n \leq 1 \tag{12.36}$$

$\Psi^{(k+1)}$ converges almost surely to Ψ^* which is the solution to the problem:

$$\min_{\Psi} C = E\left[u_t^2\right] \tag{12.37}$$

The momentum parameter ζ in (12.35) provides additional flexibility during the iterations to help achieve convergence. For example, if $C\left(\Psi^{(k+1)}\right) > \Psi^{(k)}$, so the updated parameter vector does not reduce the sum of squared errors, allowing $\zeta > 0$ acts as a method of backcasting whereby a suitable direction of the parameter vector is recast, at least partially, in terms of the previous direction of the parameter vector $\Psi^{(k)} - \Psi^{(k-1)}$.

The back propagation algorithm in (12.34) is appropriate for functions that are differentiable as is the case with the logistic function, as well as functions that are not differentiable as is the case where the threshold function in (12.13) is adopted. For the model in (12.25) with the logistic function replaced by the threshold function in (12.13), the algorithm becomes:

$$\Psi^{(k+1)} = \Psi^{(k)} + \eta \left[\begin{array}{c} \left(Y_k - Y_k^*\right) X_{1,k} \\ \left(Y_k - Y_k^*\right) X_{2,k} \end{array} \right]_{\Psi = \Psi^{(k)}} + \zeta\left(\Psi^{(k)} - \Psi^{(k-1)}\right) \tag{12.38}$$

As an example of the method of back propagation using (12.38), consider the data set given in Table 12.5.[4] The data set consists of one dependent variable Y_t, and two independent variables $X_{1,t}$ and $X_{2,t}$, defined over $T = 4$ periods. As the data are binary, the following neural network model is estimated based on the threshold function in (12.13):

$$Y_t = \left\{ \begin{array}{ll} 1: & \beta_1 X_{1,t} + \beta_2 X_{2,t} > 0.5 \\ 0: & \beta_1 X_{1,t} + \beta_2 X_{2,t} \leq 0.5 \end{array} \right. + u_t \tag{12.39}$$

The algorithm is given by (12.38) and to simplify the calculations the learning rate is chosen to be a constant $\eta = 0.2$ and the momentum parameter is set at $\zeta = 0$.

The results of estimating (12.39) by the back propagation algorithm are given in Table 12.6. The starting parameters $\Psi^{(0)} = \left[\beta_1^{(0)}, \beta_2^{(0)}\right]' = [0.1, 0.3]'$, are chosen randomly. For the first iteration, compute:

$$\begin{aligned} \beta_1^{(0)} X_{1,1} + \beta_2^{(0)} X_{2,1} &= 0.1 \times 1.0 + 0.3 \times 1.0 \\ &= 0.4 \end{aligned}$$

[4]These data are from the OR operation, a problem in symbolic logic, and taken from Medsker *et al.* (1993, p.17).

Table 12.5: Data set to demonstrate the back propagation algorithm

t	X_1	X_2	Y
1	1.0	1.0	1.0
2	0.0	1.0	1.0
3	1.0	0.0	1.0
4	0.0	0.0	0.0

which yields as the predictor from the threshold function in (12.39):

$$Y_1^* = 0.0$$

The parameter updates are computed by:

$$\begin{aligned}
\begin{bmatrix} \beta_1^{(1)} \\ \beta_2^{(1)} \end{bmatrix} &= \begin{bmatrix} \beta_1^{(0)} \\ \beta_2^{(0)} \end{bmatrix} + 0.2\,(Y_1 - Y_1^*) \begin{bmatrix} X_{1,1} \\ X_{2,1} \end{bmatrix} \\
&= \begin{bmatrix} 0.1 \\ 0.3 \end{bmatrix} + 0.2 \times (1.0 - 0.0) \begin{bmatrix} 1.0 \\ 1.0 \end{bmatrix} \\
&= \begin{bmatrix} 0.3 \\ 0.5 \end{bmatrix}
\end{aligned}$$

The calculations for the remaining iterations are given in Table 12.6. Notice that convergence is achieved after 7 iterations and the number of epochs is 2.

The algorithm is terminated when the sum of squares function is zero, that is the model provides a perfect fit. The sum of squares are reported in the final column of Table 12.6. For the starting iteration $k = 0$, the value of $\beta_1^{(0)} X_{1,t} + \beta_2^{(0)} X_{2,t}$ for each of the four periods is $\{0.4, 0.3, 0.1, 0.0\}$ which yield the corresponding predictors from the threshold function $\{0.0, 0.0, 0.0, 0.0\}$. Comparing with the actual values of $\{1.0, 1.0, 1.0, 0.0\}$ yields a value of 3.0 for the sum of squares function. The remaining values for the sum of squares are computed in a similar way.

The structure of the back propagation algorithm is for the iterations to proceed through the data set, from $t = 1, 2, ..., T$, and then repeated if need be. As the number of iterations increases, the parameters converge to that set which minimises the sum of squared errors, at least locally. It is this process, whereby the parameter estimates are improved over the data set, that in artificial neural networks is referred to as learning. This process is also referred to as training the network. Either way, it is just simply estimation which requires an iterative algorithm as a result of the nonlinear structure of the problem.

Table 12.6: Back propagation algorithm iterations using data in Table 12.5

Iteration	Epoch	t	β_1	β_2	$\sum_{t=1}^{4}(Y_t - Y_t^*)^2$
0			0.1	0.3	3.0
1	1	1	0.3	0.5	2.0
2	1	2	0.3	0.7	1.0
3	1	3	0.5	0.7	1.0
4	1	4	0.5	0.7	1.0
5	2	1	0.5	0.7	1.0
6	2	2	0.5	0.7	1.0
7	2	3	0.7	0.7	0.0

A potential problem with the method of back propagation is that convergence may be very slow, see for example, White (1988). From the discussion of gradient algorithms this is not too surprising as back propagation in essence is a steepest descent algorithm which is relatively slower than other gradient procedures using second order derivatives. To highlight the point, Table 12.7 shows the performance of the search algorithm, a gradient algorithm based on the Broyden, Fletcher, Goldfarb and Shanno algorithm (BFGS), and the method of back propagation, for estimating the parameters of the neural network:

$$Y_t = \alpha_0 + \alpha_1 Y_{t-1} + \phi_1 \left(\frac{1}{1 + \exp\left[- \left(\beta_0 + \beta_1 Y_{t-1} \right) \right]} \right) + u_t \qquad (12.40)$$

where u_t is an error term. The data generating processes are the maps given in (12.23) with $T = 250$ observations.

The search algorithm is presented based on 100, 1000 and 10000 random searches with parameters drawn from a uniform distribution in the range $[-2, 2]$. The gradient algorithm is based on the initial parameter estimates being drawn from a $N(0, 1)$ distribution. To help find the global minimum, the maximum number of initial starting draws is chosen to be 1 and 10. The back propagation algorithm is based on $\eta = 0.01t^{-0.7}$ and $\zeta = 0$ in (12.35), with a maximum of 50 epochs, and with the initial parameter estimates based on 1, 100, 1000, and 10000 random searches from the search algorithm. No other parameterisations are tried.

The results are given in Table 12.7, where in the first column the figure in parentheses indicates the number of searches or sets of starting values. These results show that the gradient algorithm performs the best in achieving the smallest residual sum of squares, while the back propagation algorithm provides a small improvement over the search algorithm. There are some gains from choosing a range of starting values to initialise the gradient algorithm.

Table 12.7: Comparison of alternative algorithms: sum of squared errors

Algorithm	Model			
	Det. chaos	Bier-Bountis	Bilinear	Stoch. chaos
Search (100)	0.000	512.165	633.247	8.618
Search (1000)	0.000	493.339	631.842	8.616
Search (10000)	0.000	492.311	631.198	8.616
Gradient (1)	0.000	343.259	647.479	8.618
Gradient (10)	0.000	343.259	629.422	8.613
Back prop. (1)	0.002	664.387	649.592	8.751
Back prop. (100)	0.000	511.835	632.311	8.618
Back prop. (1000)	0.000	493.315	631.391	8.616
Back prop. (10000)	0.000	492.311	630.774	8.616

Overall these results provide strong support for estimating artificial neural networks using standard gradient algorithms with a range of starting values. While the results of the back propagation algorithm could be improved by increasing the number of epochs and choosing alternative learning rate and momentum parameters, the gradient algorithm is also found to be computationally more efficient than the back propagation algorithm.

12.3 Testing

It is possible to perform hypothesis tests on the parameters of artificial neural networks in the same way that hypothesis tests can be performed on the parameters of econometric models. The approach is based on recognising that artificial neural networks represent approximations to the true underlying data generating process. The asymptotic distribution of the parameter estimates follows from the theory of misspecified models developed by White (1982), where the asymptotic covariance matrix of the artificial neural network estimator is given by the quasi maximum likelihood covariance estimator. White (1989) establishes the consistency and asymptotic normality of neural estimators under the conditions of *iid* data, while Kuan and White (1994) generalize these results for dependent and/or heterogeneous data. These results enable the use of Wald and Lagrange multiplier statistics for testing hypotheses.

12.3.1 Construction of Test Statistics

An advantage of using the Lagrange multiplier approach arises when the constrained model is relatively easier to estimate than the unconstrained model. This is the case in testing for neural networks. To highlight this point, consider the following model:

$$Y_t = \alpha_0 + \alpha_1 Y_{t-1} + \phi_1 \left(\frac{1}{1 + \exp\left[-\left(\beta_0 + \beta_1 Y_{t-1}\right)\right]} \right) + u_t \qquad (12.41)$$

where u_t is an error term. Under H_0, $\phi_1 = 0$, the model reduces to a linear AR(1) model. This suggests using a Lagrange multiplier statistic to test for the significance of the artificial neural network. However, there is one hurdle, namely that the parameters β_0 and β_1 are not identified under the null hypothesis. One way to circumvent this problem, which is related to the search algorithm discussed above is to treat these parameters as random; see Lee *et al.* (1993).[5]

To construct the Lagrange multiplier statistic write (12.41) as:

$$u_t = Y_t - \alpha_0 - \alpha_1 Y_{t-1} - \phi_1 \left(\frac{1}{1 + \exp\left[-\left(\beta_0 + \beta_1 Y_{t-1}\right)\right]} \right) \qquad (12.42)$$

The derivatives are:

$$\begin{aligned} Z_{1,t} &= -\frac{\partial u_t}{\partial \alpha_0} = 1 \\ Z_{2,t} &= -\frac{\partial u_t}{\partial \alpha_1} = Y_{t-1} \\ Z_{3,t} &= -\frac{\partial u_t}{\partial \phi_1} = \frac{1}{1 + \exp\left[-\left(\beta_0 + \beta_1 Y_{t-1}\right)\right]} \end{aligned} \qquad (12.43)$$

Choose the parameters $\{\beta_0, \beta_1\}$ randomly. Now evaluate the derivatives at $\phi = 0$:

$$\begin{aligned} Z_{1,t}|_{H_0} &= 1 \\ Z_{2,t}|_{H_0} &= Y_{t-1} \\ Z_{3,t}|_{H_0} &= \frac{1}{1 + \exp\left[-\left(\beta_0 + \beta_1 Y_{t-1}\right)\right]} \end{aligned} \qquad (12.44)$$

Thus the steps are:

[5] Other approaches that potentially offer greater power are by Teräsvirta and Anderson (1992) and Andrews and Ploeberger (1994). These alternative approaches are not pursued here.

1. Choose $\{\beta_0, \beta_1\}$ from a random number generator. Say $\{\widehat{\beta}_0, \widehat{\beta}_1\}$.
2. Compute:
$$F_{1,t} = \frac{1}{1 + \exp\left[-\left(\widehat{\beta}_0 + \widehat{\beta}_1 Y_{t-1}\right)\right]} \tag{12.45}$$
3. Regress Y_t on $\{1, Y_{t-1}\}$ and get the OLS residuals $\widehat{e}_t$.
4. Regress $\widehat{e}_t$ on $\{1, Y_{t-1}, F_{1,t}\}$ and compute R^2.
5. Calculate the test statistic:
$$LM = TR^2 \tag{12.46}$$
where T is the sample size, and compare with χ_1^2.

For the more general model:

$$Y_t = \alpha_0 + \sum_{i=1}^{L} \alpha_i Y_{t-i} + \sum_{i=1}^{N} \phi_i \left(\frac{1}{1 + \exp\left[-\left(\beta_{i,0} + \sum_{j=1}^{L} \beta_{i,j} Y_{t-j}\right)\right]} \right) + u_t \tag{12.47}$$

the null hypothesis is $\phi_1 = \phi_2 = ... = \phi_N = 0$. The steps for computing the LM test are:

1. Choose $\{\beta_{i,0}, \beta_{i,1}, ..., \beta_{i,L}; i = 1, 2, ..., N\}$ from a random number generator. Say $\{\widehat{\beta}_{i,0}, \widehat{\beta}_{i,1}, ..., \widehat{\beta}_{i,L}; i = 1, 2, ..., N\}$.
2. Compute:
$$F_{i,t} = \frac{1}{1 + \exp\left[-\left(\beta_{i,0} + \sum_{j=1}^{L} \beta_{i,j} Y_{t-j}\right)\right]} \tag{12.48}$$
3. Regress Y_t on $\{1, Y_{t-1}, ..., Y_{t-L}\}$ and get the OLS residuals $\widehat{e}_t$.
4. Regress $\widehat{e}_t$ on $\{1, Y_{t-1}, ..., Y_{t-L}, F_{1,t}, F_{2,t}, ..., F_{N,t}\}$ and compute R^2.
5. Calculate the test statistic:
$$LM = TR^2$$
where T is the sample size, and compare with χ_N^2.

12.3.2 Monte Carlo Results

To highlight the properties of the LM test, results of a number of Monte Carlo experiments are reported in Tables 12.8 and 12.9. The sample sizes chosen are $T = 100, 250, 500, 1000$. The number of replications is $R = 1000$.

Table 12.8 gives the results on the size of the test for two linear data generating processes:

$$AR(1) \ : \ Y_t = 0.6Y_{t-1} + v_t \tag{12.49}$$

$$AR(2) \ : \ Y_t = 0.6Y_{t-1} - 0.2Y_{t-2} + v_t \tag{12.50}$$

where $v_t \sim N(0,1)$. The neural network tests are based on the following model:

$$Y_t = \alpha_0 + \sum_{i=1}^{L} \alpha_i Y_{t-i} + \sum_{i=1}^{N} \phi_i \left(\frac{1}{1 + \exp\left[- \left(\beta_{i,0} + \sum_{j=1}^{L} \beta_{i,j} Y_{t-j} \right) \right]} \right) + u_t \tag{12.51}$$

with $N = 1, 2, 10$ activation functions, lag structures of $L = 1, 2$, and u_t is as the error term. In performing the neural network test, a generalised sweep inverse is used in the OLS regressions. This overcomes singularity problems which can occur, especially for large N.[6]

The results in Table 12.8 show that the empirical sizes are quite close to the nominal size of 5 per cent, as calculated from the asymptotic chi-square distribution, for a range of specifications of the test statistic, N and L. The results show that the asymptotic distribution approximates the finite sample distribution arbitrarily well even for sample sizes as small as $T = 100$, as most empirical sizes fall within the 95 per cent confidence interval.[7]

Table 12.9 gives the results on the power of the neural network test assuming that the models under the alternative hypothesis are the bilinear and the stochastic chaos models in (12.23). The bilinear model is chosen for comparison with the Monte Carlo results of Lee *et al.* (1993). The chaos model allows for a stochastic parameter to vary between 0 and 4, thereby allowing the data generating process to exhibit a rich array of characteristics. The results of the Monte Carlo experiments for the bilinear model show that the power of the neural network test in general increases as the number of activation functions N increases, but decreases as the lag length L is increased. For the stochastic chaos model, the power of the neural network test is extremely high for all parameterisations of the test statistic.

---

[6] Lee *et al.* (1993) suggest using principal components to circumvent potential singularity problems.

[7] Letting $s = 0.05$ be the size of the test, a 95 per cent confidence interval is given by $\pm 1.96\sqrt{s(1-s)/R} = (3.6, 6.4)$ where $R = 1000$ is the number of replications.

Table 12.8: Size (percentage) of neural network tests: 5 per cent significance level

Parameters	T	AR(1)	AR(2)
	100	3.90	2.00
$N = 1$	250	3.40	2.10
$L = 1$	500	4.60	2.50
	1000	4.70	1.90
	100	3.10	2.20
$N = 2$	250	3.50	2.40
$L = 1$	500	3.30	1.20
	1000	5.20	2.70
	100	3.60	3.40
$N = 10$	250	3.80	3.30
$L = 1$	500	2.60	3.30
	1000	4.20	3.10
	100	5.20	6.50
$N = 1$	250	5.90	5.70
$L = 2$	500	5.00	5.20
	1000	4.80	4.50
	100	5.60	5.30
$N = 2$	250	4.80	4.50
$L = 2$	500	5.10	5.20
	1000	5.40	4.40
	100	3.40	4.20
$N = 10$	250	4.10	4.40
$L = 2$	500	3.70	4.40
	1000	5.00	5.80

Table 12.9: Power (percentage) of neural network tests[a]

Parameters	T	Bilinear	Stoch. Chaos
	100	59.10	97.90
N=1	250	72.60	99.00
L=1	500	80.70	99.00
	1000	85.00	99.20
	100	85.50	99.90
N=2	250	94.50	100.00
L=1	500	98.90	100.00
	1000	99.10	100.00
	100	93.80	97.50
N=10	250	99.90	99.50
L=1	500	99.90	99.60
	1000	100.00	99.80
	100	29.70	66.70
N=1	250	42.70	80.00
L=2	500	52.10	86.00
	1000	61.00	91.00
	100	49.20	88.40
N=2	250	67.00	94.30
L=2	500	77.00	97.10
	1000	85.50	99.80
	100	99.10	99.40
N=10	250	100.00	100.00
L=2	500	100.00	100.00
	1000	100.00	100.00

(a) Based on 5 per cent (asymptotic) significance level.

12.4 Relationships with Econometric Models

The analysis has so far focused on showing the relationships between artificial neural networks and a broad class of econometric models. In this section, special attention is given to showing specific relationships with some nonlinear time series models and binary choice models.

12.4.1 Nonlinear Time Series Models

Many of the recent nonlinear time series models represent neural networks with different activation functions. Two examples are the self-exciting autoregressive threshold class of models (SETAR), discussed in detail by Tong (1990) and the smooth transitional autoregressive model (STAR) used by Teräsvirta and Anderson (1992) in estimating nonlinear models of international business cycles. With one lag, the SETAR model is:

$$Y_t = \alpha_0 + \alpha_1 Y_{t-1} + \phi_1 Y_{t-1} D_t + u_t \tag{12.52}$$

where u_t is an error term, and D_t is a dummy variable given by:

$$D_t = \begin{cases} 1: & Y_{t-\tau} > \xi \\ 0: & Y_{t-\tau} \leq \xi \end{cases} \tag{12.53}$$

The parameters ξ and τ are referred to as the threshold and delay parameters respectively. This model is an artificial neural network with a threshold activation function as given in (12.13). For the STAR model, the threshold activation function is replaced by the logistic activation function (12.11):

$$D_t = \frac{1}{1 + \exp\left[-\left(\beta_0 + \beta_1 Y_{t-\tau}\right)\right]} \tag{12.54}$$

12.4.2 Binary Choice Models

Binary choice models can be interpreted as neural networks. This is important as it provides insights into how existing binary choice models can be generalised. For examples of other ways of generalising binary choice models, see chapter 4.

Let Y_t be a binary random variable with the property that:

$$Y_t = \begin{cases} 1: & P_t > 0.0 \\ 0: & P_t \leq 0.0 \end{cases} \tag{12.55}$$

where P_t is the probability that $Y_t = 1$. Choosing P_t as the logistic distribution:

$$P_t = \frac{1}{1 + \exp\left[-\left(\beta_0 + \beta_1 X_t\right)\right]} \tag{12.56}$$

and X_t as a set of attributes of the tth individual, gives the logit model. As P_t represents the conditional mean of Y_t, then in neural network terminology the output variable is:

$$Y_t^* = E\left[Y_t | X_t\right] = P_t \tag{12.57}$$

where the activation function is the logistic function.

The relationship between artificial neural networks and the logit model suggests how qualitative response models can be generalised to allow for more flexible functional forms; see Amemiya (1981). As an example, consider the probit model where P_t in (12.55) is the cumulative normal distribution, F_N. A generalisation of the probit model is to choose:

$$P_t = F_N\left(\beta_0 + \beta_1 X_t + \frac{\phi}{1 + \exp\left[-\left(\gamma_0 + \gamma_1 X_t\right)\right]}\right) \tag{12.58}$$

where the index of attributes $\beta_0 + \beta_1 X_t$, is augmented by the nonlinear logistic activation function. Keeping in the spirit of qualitative response models, this model could be referred to as to neurit. A special case of this model is the probit model occurring at $\phi = 0$. This suggests that a Lagrange multiplier test of normality can be constructed by testing the restriction $\phi = 0$.

12.5 An ANN Foreign Trading System

There is growing interest in applying ANNs to forecasting exchange rates and other financial variables; see for example White (1988), Pictet *et al.* (1992), Weigend *et al.* (1992). This research has tended to concentrate on the predictive properties of ANNs with little attention given to incorporating the forecasts into a trading system which explicitly takes into account transaction costs; see also chapter 11 for the development of a trading system using genetic algorithms. The role of transaction costs is important as the potential profit opportunities arising from alternative forecasting models may be overstated if transaction costs are ignored.

This application has a dual role: first, to investigate the forecasting properties of an ANN and compare it with a linear AR model; and secondly, to construct a trading model which takes into account transaction costs.[8]

12.5.1 The Trading System

The aim of the foreign exchange system is to identify movements of the exchange rate which are sufficiently large, that is in excess of transaction

[8]There are few papers which include transaction costs when evaluating the performance of alternative forecasting models; one example is Brock *et al.* (1992).

costs, which in turn generate profits. The foreign exchange trading system is summarized as:

$$e_{t+1}^f - e_t \begin{cases} > U_t + H & : \text{ Buy} \\ < L_t - H & : \text{ Sell} \\ \text{else} & : \text{ Do nothing} \end{cases} \tag{12.59}$$

where e_t is the exchange rate at the time a strategy is being devised, e_{t+1}^f is the forecast of the exchange rate in the next period, U_t and L_t are respectively the upper and lower boundaries determined by the arbitrage boundaries set by the interest differential, and H is a filter designed to eliminate unprofitable trades by filtering out small moves forecast in the exchange rate. In the empirical application this filter is allowed to vary from $H = 0.0000$ to $H = 0.0020$ in increments of 0.0005.

The interest differential in terms of foreign exchange points in deciding whether to buy foreign currency equals the foreign asset deposit interest rate less the domestic funding cost. Thus the upper boundary is computed as:

$$U_t = \left[\frac{1}{\left(e_t + \frac{\varphi}{2}\right)}\left(\frac{1 + r_t^* - \frac{\varsigma}{2}}{52}\right)\left(e_{t+1}^f - \frac{\varphi}{2}\right)\right] - \left[\frac{1 + r_t + \frac{\varsigma}{2}}{52}\right] \tag{12.60}$$

where r_t and r_t^* are respectively the domestic and foreign interest rates, and $\varphi = 0.0007$ and $\varsigma = 0.0002$ respectively represent the foreign market and money market transaction costs.[9] The interest differential in terms of foreign exchange points in deciding whether to sell foreign currency equals the domestic asset deposit interest rate less the foreign funding cost. Thus the lower boundary is computed as:

$$L_t = \left[\frac{1 + r_t - \frac{\varsigma}{2}}{52}\right] - \left[\frac{1}{\left(e_t - \frac{\varphi}{2}\right)}\left(\frac{1 + r_t^* + \frac{\varsigma}{2}}{52}\right)\left(e_{t+1}^f + \frac{\varphi}{2}\right)\right] \tag{12.61}$$

If the trading system generates a buy signal, the position is closed the next week by selling the foreign currency at the current closing rate. The opposite occurs when the system generates a sell signal.

12.5.2 Data and Specification

The data consist of the closing price of the Australian/US dollar in Sydney, e_t, the closing Australian cash rate in Sydney, r_t, and the closing US Federal Funds rate in New York, r_t^*. Thus Australia represents the domestic country

[9] The transaction cost parameters ς and φ in (12.60) and (12.61) are determined from the bid-ask spreads in the foreign and money markets respectively.

and the US is the foreign country. The data are weekly beginning on 1 January 1986 and ending on 14 June 1995, a total of 495 observations. All data are obtained from the Reserve Bank of Australia.

A range of ANNs are estimated using the back propagation algorithm with over 25,000 iterations. The parameters of this algorithm in (12.35) are set at $\eta = 0.07$ and $\zeta = 0.1$. Input noise of 1 per cent is added to the parameter estimates to help find the global minimum. The training tolerance is set at 0.01.

The estimation period, that is the training period, is the first 469 observations of the data set. The next five observations are used to decide when to stop the algorithm, while the remaining 21 obervations are used for out-of-sample forecasting.

The model chosen has three hidden layers, one output and three lags:

$$e_t = \sum_{i=1}^{3} \phi_i F_{i,t} + u_t \tag{12.62}$$

where u_t is an error term and the hidden layers are given by the logistic function:

$$F_{i,t} = \frac{1}{1 + \exp\left[-\left(\beta_{i,0} + \sum_{j=1}^{3} \beta_{i,j} e_{t-j}\right)\right]} \tag{12.63}$$

12.5.3 Results

The profits earned by the ANN trading system are calculated for the 21 weeks of the out-of-sample period. The profits and losses are given in terms of foreign exchange points in local currency (Australian dollar) terms. For example, a profit of 0.0500 points is equivalent to 5 cents for every Australian dollar traded or 5 per cent of the traded amount. For comparison, the performance of the ANN model is compared with the performances of a perfect foresight model and a linear AR(2) model. The perfect foresight model assumes that every single trade is correctly executed so that all profitable trades are executed. The AR model is estimated over the first $474 = 469 + 5$ observations. Other lag structures were tried but were found to be inferior in terms of forecast errors.

Table 12.10 gives profits and losses of the three alternative models for various filter values of H given in (12.59). The perfect foresight results show that the maximum profit earned over the period is just under 20 cents per Australian dollar traded. The best return using the ANN to forecast the exchange rate is 5.46 cents per Australian dollar traded , which occurs for a filter of $H = 0.0000$. This translates into 14.2 ($=5.46 \times 52/20$) cents per annum, which is in excess of the risk free rate of interest over this period.

Table 12.10: Comparison of alternative model's profits

Filter	Model		
	Perfect foresight	ANN	AR(2)
$H = 0.0000$	0.1792	0.0546	-0.0074
$H = 0.0005$	0.1792	0.0220	-0.0074
$H = 0.0010$	0.1776	0.0110	-0.0074
$H = 0.0015$	0.1765	0.0110	-0.0184
$H = 0.0020$	0.1763	0.0138	0.0163

Although the ANN model does not generate profits close to those obtained by the perfect foresight model, it is superior to the AR model which generates losses for all filter values with the exception of $H = 0.0020$.

Table 12.11 provides a breakdown of the number of winning trades for the alternative forecasting models. Of the 21 weeks, from the perfect forecasting results the optimal strategy is to buy US dollars for 12 periods, sell for 8 periods and do nothing for 1 period. For a filter of $H = 0.0000$, the ANN model gets half the number of winning trades correct. It is interesting to note that the AR model gets one fewer winning trade than the ANN model whilst from Table 12.10 its profit performance is inferior to the ANN model. A further breakdown of the results shown in Table 12.11 reveals that the average profit earned over all trades for the ANN model is larger than it is for the AR model, whereas the maximum loss incurred for the ANN model is smaller than it is for the AR model.

12.6 Conclusions

This chapter has provided an introduction to artificial neural networks, adopting an econometric approach whereby artificial neural networks are analysed in terms of model specification, estimation and testing. The key result is that much of the neural network literature has a direct parallel in econometrics and in statistics more broadly. Special attention was given to investigating the properties of alternative estimation algorithms, known as learning in neural network parlance, and the properties of neural network tests. In the case of estimation methods, standard gradient algorithms are found to be superior to the back propagation algorithm, the method commonly adopted in the neural network literature, in terms of locating the global minimum and also in terms of computational efficiency. As regards hypothesis testing, the neural network test provides an important diagnostic for testing against alternative forms of nonlinearity.

Table 12.11: A break down of trades for alternative models

Performance (number of trades)	Model	Filter (T) 0.0000	0.0005	0.0010	0.0015	0.0020
Winning trades	PF	20	20	18	17	17
	ANN	10	9	8	8	8
	AR	9	9	9	8	8
Buy	PF	12	12	11	11	11
	ANN	6	5	5	5	5
	AR	6	6	6	6	6
Sell	PF	8	8	7	6	6
	ANN	13	13	12	12	11
	AR	12	12	12	11	10
Do nothing	PF	1	1	3	4	4
	ANN	2	2	4	4	5
	AR	3	3	3	4	5

An artificial neural network foreign exchange trading model was estimated and the profits compared with a perfect foresight model and a linear autoregressive model. The neural network model generated profits in excess of a risk-free rate and, in particular, generated profits in excess of the AR model, which were found to be negative in general. Comparing the results to a perfect foresight model revealed that there are profitable gains from conducting further research in this area.

Chapter 13

An ANN Model of the Stock Market

G.C. Lim and P.D. McNelis

This chapter examines the properties of an artificial neural network (ANN) model of daily stock price returns in Australia which includes the responses to shocks in foreign markets. It considers the question of whether Australian stock returns are predictable on a daily basis, and if so, whether the Japanese stock market matters more than the US stock market for making such predictions. The ANN model is compared to an autoregressive linear model and a GARCH-M model.

The relative importance of the US and Japanese economy in Australasian financial markets has been examined by Frankel (1991) and Frankel and Wei (1993) in the context of interest and exchange rate linkages. They find that the Asian-Pacific regions remain very much a US-dominated area, despite the growing integration of financial markets around the Pacific Rim. This chapter extends the study of the relative influence of the Japanese and US markets in the Pacific Rim to equity markets.

Section 13.1 discusses the statistical properties of the three indices used: the Australian All-Ordinaries (AllOrds), the Japanese Nikkei index and the US Standard and Poor (S&P) index. The high-frequency data was also subjected to the Gallant-Rossi-Tauchen filtering to correct for calendar – daily, weekly, monthly and holiday – effects. The main result obtained from a contemporaneous correlation analysis is that, in log-level terms, the AllOrds is correlated more with the S&P than with the Nikkei, but in log-differenced terms (a measure of daily returns), the AllOrds is correlated more with the Nikkei than with the S&P.

Section 13.3 presents estimates of the linear autoregressive model. When lagged adjustment is allowed, the results show that returns in the AllOrds are more affected by shocks in the S&P Index than by shocks in the Nikkei Index. Section 13.3 extends the analysis to the GARCH-M model which allows for nonlinearities via a time-varying variance in the daily stock returns. These results are similar to those obtained for the linear model with constant volatility.

Section 13.4 applies the neural network analysis to evaluating how well shocks to the S&P and to the Nikkei indices influence returns in the AllOrds. Nonlinearities are incorporated in the form of the parallel processing of data, but since neural networks are shortcuts to more complicated nonlinear system analysis, several neural architectures are explored. The preferred model is a two-neuron neural network model based on a time-zone trading framework – one neuron captures the processing of same-day information (the information contained in shocks to the Nikkei index, the Pacific time zone effect, may be exploited in the Australian market that same-day); and another neuron captures the processing of next-day information (information contained in shocks to the S&P index, the Atlantic time-zone effect, may only be exploited by Australian traders next day when the Australian market reopens).

The results show that daily returns in the Australian stock index are predictable, and in rank orderings, the neural network model generates, in-sample, the lowest residual sum of squared errors, followed by the autoregressive model and then the GARCH-M model. A comparison of the estimated results show that, relative to the nonlinear neural network model, the linear autoregressive model underestimates the effect of the Japanese stock market and overestimates the effect of the US stock market on Australian stock returns.

For the out-of-sample forecasting analysis, the restricted feedforward ANN model, incorporating parallel processing of information according to an East-West specification, out-performs the linear model, as well as other types of neural architecture, in having the lowest root mean squared errors.

13.1 Data: Statistical Analysis and Filtering

Figure 13.1 shows the evolution of the AllOrds, Nikkei, and S&P indices, in log form. The data are daily from 28 September 1990 to 27 September 1995 (1305 observations); for convenience of comparison, the indices have been standardised to the same base. Figure 13.1, for data denominated in their home currency, suggests that the AllOrds contains an upward trend which is positively correlated (0.845) with the S&P, as well as a business cycle component which is negatively correlated (-0.410) with the Nikkei. The correlation between the Nikkei and the S&P is -0.668.

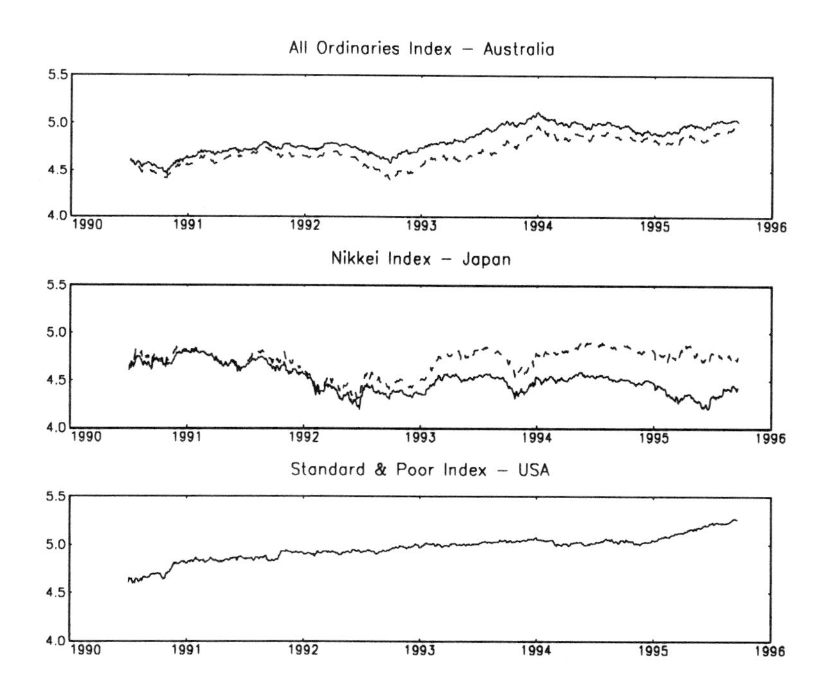

Figure 13.1: Stock indices (logs): in home currency (solid), in $US (dashed)

However, comparing these indices denominated in their home currency may be misleading because no allowance is made for the effect of exchange rate variations. There has been a continual appreciation of the Yen over this period, so it would be desirable to remove the effects of exchange rate movements. To abstract from variations in the exchange rates, Figure 13.1 also shows the stock indices expressed in $US equivalents, using daily exchange rates. The most significant point to note is that the correlations between the AllOrds and the Nikkei switches from negative (-0.410) to positive (0.503) when the effects of the exchange rate are taken into account. The correlation between the AllOrds and the S&P is still positive 0.733), while the correlation between the Nikkei and the S&P is now 0.166. The positive relationships between all three $US-based indices reflect the tendency of financial assets to move in line with each other. But, from an analysis of the contemporaneous correlations between the logs of these $US-based indices, it is noted that the co-movements between the S&P and the Nikkei is weak, whereas that between the AllOrds and the Nikkei and S&P are stronger, with the trend in the AllOrds being more influenced by movements in the S&P than by the Nikkei.

Consider the daily returns, represented by the log-differences expressed in percentage form, shown in Table 13.1. The correlation between the AllOrds and the Nikkei is twice as large as that between the AllOrds and the S&P. Thus, preliminary contemporaneous analysis of the data suggests that while the trend in the AllOrds is more highly correlated with the S&P, the daily returns of the AllOrds show a stronger positive relationship with the log-differences of the Nikkei.

Finally, multicountry analysis of stock market data requires special treatment for high-frequency (daily) data. Despite a relatively large number of observations, holiday effects, which differ across countries, may distort the outcomes. Several combinations of days with no trading in New York and much price movement in Australia, or days with no trading in Australia, when there may be movement in New York, can put a downward bias on the true effect of New York movements on Australian stock prices. Similarly, there may be common calendar effects across countries: heated trading days during the middle of the business week, and diminished trading during holiday periods in December and January, which can put an upward bias on a statistical analysis of the interrelations. Thus, a careful pre-filtering of the data is an important step.

Gallant *et al.* (1992) have offered a special filtering method. Each stock index series, in logarithmic first differences, is regressed against the following independent variables: dummy variables for each month of the year from February through November; special dummies for each week in December and January; daily dummy variables for Tuesday through Friday; and finally

Table 13.1: Descriptive statistics of the indices (base year = 100, logs)

	Indices in home currency			Indices in $US		
	AllOrds	Nikkei	S&P	AllOrds	Nikkei	S&P
mean	4.824	4.512	4.964	4.703	4.705	4.964
std. dev.	0.154	0.145	0.131	0.141	0.137	0.131
Correlations between indices						
	1.000			1.000		
	-0.410	1.000		0.503	1.000	
	0.845	-0.668	1.000	0.733	0.166	1.000
	Log-Differences: ($US-based Indices)×100					
	Unfiltered data (1266 obs)			Filtered data (1232 obs)		
	AllOrds	Nikkei	S&P	AllOrds	Nikkei	S&P
mean	0.025	0.011	0.050	0.024	0.010	0.048
std. dev.	0.959	1.629	0.674	0.952	1.622	0.665
Correlations between Indices						
	1.000			1.000		
	0.193	1.000		0.176	1.000	
	0.073	0.110	1.000	0.080	0.103	1.000

dummy variables set at the square root of the number of elapsed days since the last trading day. Normally, this implies that each Monday is set at the square root of two, since the preceding Saturday and Sunday are nontrading days.

For Australia, each holiday is also deleted from the sample, resulting in 72 observation being eliminated. But having deleted, for example, the Queen's Birthday holiday, which falls on a Monday, the calendar dummy is set at the square root of three for the following Tuesday. For the US and Japan, holidays in these countries that are not holidays in Australia are not deleted from the sample. A nontrading day on, for example, Friday, 4 July, a holiday in the US, will simply imply filtering by a holiday dummy which is the square root of three for the following Monday. The holidays for the other markets which are not holidays in Australia were not deleted because the focus of the analysis is Australia: the aim is to explain as much as possible of the movements in the All-Ordinaries. Eliminating days of price movements in Australia, due to a holiday in the US or in Japan, would eliminate too much information from the sample. While leaving in holidays in the US and Japanese series, with trading days in Australia, may downwardly bias any statistical relationships, this potential problem is circumvented somewhat by the use of a calendar variable in the filtering of the S&P and the Nikkei.

The last step in filtering the data is to scale the residuals from the calendar regression so that the filtered series have the same mean and standard deviation as the original log-differenced series. The data so generated is the filtered data now purged of the calendar day-of-the week, month-of-the-year and holiday effects.

A statistical analysis of the filtered data is also contained in Table 13.1. Since the correlation results based on filtered data are similar to the correlation results based on the raw data, the calendar effects are not distortionary. This result is not surprising since all three markets under study have developed sophisticated financial systems.

13.2 The Linear Autoregressive Model

The model typically applied to extend contemporaneous correlation analysis to allow for lagged effects is the linear dynamic autoregressive model:

$$y_t = \alpha_0 + \sum_{i=1}^{p} \alpha_i y_{t-i} + \sum_{j=0}^{q} \beta_j x_{t-j} + \sum_{k=1}^{r} \gamma_k z_{t-k} + \varepsilon_t \tag{13.1}$$

$$\varepsilon_t \sim N(0, \sigma^2) \tag{13.2}$$

where y is the All-Ordinaries index, x is the Nikkei, z is the S&P and ε is the error term with zero mean and constant variance. All the data are in log-differences. Also, since the Australian market is small relative to the New York and Tokyo stock exchanges, the analysis treats the S&P and Nikkei indices as exogenously determined to the AllOrds. Two empirical questions have to be addressed. The first is whether to include current values of the Nikkei and S&P in the set of regressors, and the second is the order of the lags.

With respect to the first question, current values of the Nikkei are included, but only lagged values of the S&P. This is because there is only an hour difference between Australian Eastern standard time (AEST) and the time in Tokyo; hence current information about the Japanese stock market would be available to Australian dealers within a day's trading. However, since the time difference between AEST and New York is almost a day, this implies that current US information can only be exploited the next day in the Australian stock exchange.

With respect to the order of lags, the Akaike, Schwarz and Hannan-Quinn criteria were adopted to discriminate between alternative models. Using the same model with increasingly longer lags, the optimal number of lags was found to be $p = r = 2$ and $q = 1$.

Table 13.2: Linear regression model: filtered data

Variable		Coefficient		*t*-statistic
Constant		0.007		(0.278)
AllOrds(-1)		0.111		(3.880)
AllOrds(-2)		-0.065		(-2.450)
Nikkei		0.073		(4.480)
Nikkei(-1)		-0.004		(-0.267)
S&P(-1)		0.431		(9.870)
S&P(-2)		-0.081		(-2.004)
R^2	=	0.138		
Residual sum of squares	=	959.220		
AIC	=	8.452		
SIC	=	8.488		
HIC	=	8.452		
Residual diagnostics: p-values in brackets				
AR(1)	=	0.000	(0.999)	
ARCH(1)	=	6.500	(0.011)	
JB test for normality	=	25.440	(0.000)	
LWG test for nonlinearity	=	7.399	(0.025)	

The results from the estimation of equation (13.1) with $p = r = 2$ and $q = 1$ are presented in Table 13.2 for the filtered data. Since the data series are stationary, the significance of the coefficients for each country can be evaluated by standard econometric techniques. Table 13.2 shows that with this linear model, the Australian own lags, the lagged returns on the S&P and the current return on the Nikkei are significant explanatory variables, at the 5 per cent level, of returns in the All-Ordinaries. However, the S&P has the greatest effect.

The problem with the classical linear approach to the analysis of stock-price movements is that it ignores or sidesteps asymmetries in the data. Slow upward movements, followed by quick declines, indicating the bursting of a bubble, suggests non-linear or at least a different statistical processes than those implied by equation (13.1) with a constant variance. As can be seen from the residual diagnostics, both the ARCH test and the neural network test of Lee et al (1993) indicate the presence of non-linearities in the system. The LWG test is designed to determine whether hidden units to the linear network would be advantages; the result show a rejection of the null of linearity.

13.3 The GARCH-M Model

Generalised autoregressive conditional heteroscedastic (GARCH) models attempt to capture a nonlinear component in the stochastic process by allowing the second moment to be time-varying. The GARCH-in-mean model is represented by the following system, originally due to Bollerslev (1986):

$$y_t = \alpha_0 + \sum_{i=1}^{p} \alpha_i y_{t-i} + \sum_{j=0}^{q} \beta_j x_{t-j} + \sum_{k=1}^{r} \gamma_k z_{t-k} + \theta\sqrt{h_t} + \varepsilon_t \tag{13.3}$$

$$\varepsilon_t \sim N(0, h_t) \tag{13.4}$$

$$h_t = \delta_0 + \delta_1 \varepsilon_{t-1}^2 + \delta_2 h_{t-1} \tag{13.5}$$

where h is the time-varying variance. The coefficients in Table 13.3 are estimated by maximum-likelihood techniques for the same lag structure as the linear dynamic model given in equation (13.1) using EVIEWS, and the t-statistics are based on Bollerslev-Woolridge robust standard errors and covariances.

The GARCH-M estimation produces results similar to the linear autoregressive model and, while the GARCH-M parameter, θ, is positive, it is not significant. However, δ_0, δ_1, and δ_2 are significant, and as shown later, a model allowing for time-varying volatility can out-perform the linear model in out-of-sample forecasting for certain days forward.

13.4 The Neural Network Model

From chapter 12, the neural network model with n neurons and one hidden layer is:

$$y_t = \phi_0 + \sum_{n} \phi_n N_{nt} \tag{13.6}$$

$$N_{nt} = (1 + e^{-\lambda_{nt}})^{-1} \tag{13.7}$$

$$\lambda_{nt} = \omega_{n0} + \sum_{m} \omega_{nm} I_t \tag{13.8}$$

At time t, the information from the m inputs, $\{y_{t-i},\ x_{t-j},\ z_{t-k}\}$, is aggregated with weights ω_{nm}, where the bias or constant factor is ω_{n0} in equation (13.8), and the information is then passed through an activation or threshold

Table 13.3: The GARCH-M model

		Coefficient	t-statistic
Constant		-0.377	(-0.845)
AllOrds(-1)		0.119	(3.961)
AllOrds(-2)		-0.065	(-2.437)
Nikkei		0.072	(4.138)
Nikkei(-1)		0.004	(-0.217)
S&P(-1)		0.429	(10.210)
S&P(-2)		-0.092	(-2.306)
GARCH-M parameters			
θ		0.441	(0.874)
δ_0		0.522	(2.326)
δ_1		0.084	(1.984)
δ_2		0.249	(0.824)
R^2	=	0.139	
Residual sum of squares	=	958.530	
AIC	=	8.460	
SIC	=	8.515	
HIC	=	8.459	

function, equation (13.7), to generate a value N_{nt}. Each neuron N_{nt} may be interpreted as a 'squasher' of information and the activation function is usually nonlinear, typically the log-sigmoid function because it is differentiable and hence convenient for the back-propagation algorithm to be discussed later. The n neurons are in turn passed through to output y_t in a linear form with weights ϕ_n and bias or constant factor ϕ_0; equation (13.6).

A limitation of neural network models is that they require a large number of parameters. In the case of n neurons and k inputs, the required number of parameter estimates is $(n + 1) \ + n(k + 1)$. Naturally, the more neurons and inputs, the better the fit and forecasting power, but the model can be over-parameterised. In fact, with a large set of weights, the network can be trained so that the sum of squared residuals is minimised at zero.

But for researchers who look for simplicity and parsimony, the neural network approach has too many free parameters. Too many different outcomes are consistent with learning. As yet, there are not enough restrictions, both on the number of neurons and on the functional form of the neurons, on the way the neurons react to input stimuli, and on the way the neurons affect output. Neural networks represent a line of research with obvious appeal to economic modellers, but await further theoretical development.

This chapter is less concerned with the forecasting potential of neural networks than with finding out if there is a useful way of characterising input information and neurons. The analysis is restricted to a neural system with only one hidden layer and two neurons. Four neural architectures were examined and the results are presented in Table 13.4. The models were estimated with the learning parameter set at 0.3 and the momentum parameter set at 0.5. The number of epochs was set at 30,000.

Unrestricted model

This is the most general form of a neural network analysis: the same set of inputs I_t feeds into all neurons. Results in Table 13.4, model 1, show that it has a lower residual sum of squares than the linear autoregressive model.

Domestic/Foreign classification

To make economic sense of a neural network model for financial analysis, beyond pure predictive accuracy, it would be useful to characterise the input information according to some economic criteria. For example, information could be processed according to whether it is a domestic or a foreign factor. Thus neuron 1 could capture information from lagged AllOrds, while neuron 2 could 'squash' information contained in the current and lagged Nikkei and the lagged S&P. The result for this specification is in Table 13.4, model 2,

Table 13.4: Alternative models: minimum residual sum of squares

Model 1	Unrestricted form	941.475
N_1	AllOrds(-1), AllOrds(-2), S&P(-1), S&P(-2) Nikkei, Nikkei(-1)	
N_2	AllOrds(-1), AllOrds(-2), S&P(-1), S&P(-2) Nikkei, Nikkei(-1)	
Model 2	Domestic/Foreign	958.359
N_1:	AllOrds(-1), AllOrds(-2)	
N_2:	S&P(-1), S&P(-2), Nikkei, Nikkei(-1)	
Model 3	East/West	945.220
N_1:	AllOrds(-1), AllOrds(-2), Nikkei, Nikkei(-1)	
N_2:	AllOrds(-1), AllOrds(-2), S&P(-1), S&P(-2)	
Model 4	Atlantic/Pacific time zones	940.934
N_1:	AllOrds(-1), AllOrds(-2), S&P(-1), S&P(-2) Nikkei(-1)	
N_2:	AllOrds(-1), Allords(-2), S&P(-1), S&P(-2) Nikkei(-1), Nikkei	

and it shows that this characterisation is only marginally better than the linear autoregressive case.

East/West geographic model

A better way to characterise the information is to take into account a geographical fact that one set of information is coming from Japan (the East) while other set of information is coming from the US (the West). The results in Table 13.4, model 3, show a marked improvement in the residual sum of squares relative to the unrestricted case.

Pacific/Atlantic time zone model

Finally, a restricted feedforward neural network model, which distinguishes between current and lagged information structures, is examined. This neural architecture is based on a time-zone classification which recognises that signals are transmitted to the AllOrds from two different groups of traders: those with access to same-day information about the Nikkei index (Pacific time-zone traders), and those with access to information for making decisions with at least a one-day lag (for the sake of simplicity, Atlantic time-zone traders). The results are in Table 13.4, model 4, and show that this model performs best in having the least residual sum of squares. Note that

Table 13.5: Feedforward neural network: time-zone neural architecture

Input layer weights		Neuron 1	Neuron 2
Bias		2.401	-0.812
AllOrds(-1)		0.076	0.399
AllOrds(-2)		0.044	-0.714
Nikkei		0.115	0.000
Nikkei(-1)		-0.189	0.947
S&P(-1)		0.420	1.327
S&P(-2)		0.161	-1.443
Hidden layer weights			
Bias		-7.555	
Neuron 1		7.908	
Neuron 2		0.986	
R^2	=	0.155	
Residual sum of squares	=	940.934	
AIC	=	8.440	
SIC	=	8.486	
HIC	=	8.439	
Residual diagnostics: p-values in brackets			
AR(1)	=	1.459	(0.227)
ARCH(1)	=	5.848	(0.016)
JB test for normality	=	43.681	(0.000)

since the model is nonlinear, the explanatory power of the model has actually improved, compared to the unrestricted case, with one less parameter.

Since model 4 is the preferred specification, this model's weights are presented in Table 13.5. A comparison with the results from the linear model, Table 13.2, shows that the nonlinear model produces the lower sum of squared residuals. Moreover, according to the model selection information criteria, the Akaike, Schwarz and Hannan-Quinn, the restricted feedforward neural network architecture model is preferred against the linear model.

Diagnostic tests of the residuals, however, show similar results. Both sets of error distribution are non-normal, and although free of autocorrelation, an ARCH(1) process could only be rejected at the 1 per cent level.

Finally, for a comparison of the relative effects of shocks to the Nikkei and the S&P on the AllOrds, the dynamic multipliers for a unit change in the input variables are computed. These results are presented in Table 13.6. The most important point to note is that the linear model, in comparison to the nonlinear model, understates the effects of shocks to the Nikkei and

Table 13.6: Dynamic multipliers: relative to samples means

Input variables			
Linear autoregressive model			
Period	AllOrds(-1)	Nikkei	S&P(-1)
1	0.111	0.073	0.433
2	-0.053	0.080	0.397
3	-0.013	0.076	0.365
4	0.002	0.075	0.364
5	0.001	0.075	0.366
Nonlinear neural network model			
Period	AllOrds(-1)	Nikkei	S&P(-1)
1	0.134	0.066	0.527
2	-0.087	0.186	0.310
3	-0.027	0.194	0.221
4	0.007	0.178	0.237
5	0.004	0.174	0.249

overstates the effects of shocks to the S&P on returns on the AllOrds. The nonlinear results also suggest that, while the impact effect of shocks in the S&P dominates the impact effect of shocks to the Nikkei, over time the effects of the S&P decline, while those of the Nikkei increase.

13.5 Out-of-sample Forecasting

Table 13.7 shows the root mean squared errors for a number of days forward. The results show that, excluding the GARCH-in-mean model, the East/West model consistently performs best out-of-sample. This is in contrast to the time-zone model which performed best within-sample, suggesting that there are forecasting gains in specifying an appropriate neural architecture.

The good performance of the GARCH model for 3-5 days forward should be treated with some caution, as it is not a consistent result, and hence may reflect the particular sample considered. Nevertheless, it would appear that there are forecasting gains to modelling volatility.

13.6 Conclusions

This chapter has used a nonlinear neural network model to examine asset-price developments. The Pacific/Atlantic time zone neural architecture for a

Table 13.7: Root mean squared errors: out-of-sample forecasts

Days Forward	Model					
	OLS	GARCH	Neural unrest.	Domestic/ foreign	East/ West	Atlantic/ Pacific
1	0.312	0.268	0.280	0.308	0.240*	0.266
2	0.221	0.232	0.198	0.218	0.181*	0.188
3	0.349	0.203*	0.319	0.347	0.288	0.315
4	0.375	0.275*	0.417	0.379	0.319	0.425
5	0.509	0.330*	0.516	0.514	0.453	0.511
10	0.499	0.528	0.530	0.504	0.466*	0.528
20	0.577	0.714	0.576	0.577	0.559*	0.574
40	0.761	0.850	0.762	0.763	0.752*	0.771
60	0.726	0.855	0.728	0.726	0.717*	0.738
80	0.706	0.795	0.703	0.705	0.701*	0.714

* Denotes lowest root-mean-square errors.

single-layer, two-neuron network offers a parsimonious strategy for incorporating nonlinearities and has the advantage of being based on economically coherent information restrictions. This model can be extended to incorporate European time-zone traders, with partial access to Pacific and Atlantic time-zone information.

The results also indicate that ARCH processes remain in the residuals. Thus nonlinearities arising from GARCH processes, which feed back into time-varying risk premia, may complement the model and improve the predictability of stock price movements.

Chapter 14

Exchange Rate Forecasting Models

Vance L. Martin, Eugene S.Y. Choo and Leslie Teo

Forecasting exchange rates is an extremely difficult task facing applied econometricians. The difficulties involved are best highlighted by the work of Meese and Rogoff (1983) who find that no structural model delivers superior forecasts to the simplest of all models, the random walk. Although this result is supportive of the efficient-markets theory (Baillie and McMahon, 1990), there is a general feeling of uneasiness.

Whilst most applied studies find little forecasting power in economic models, there are some exceptions. Boughton (1987), in using a preferred habitat model where assets are imperfect substitutes, finds for certain periods large improvements in forecastability over the random walk. Fisher *et al.* (1990) also find that improvements in forecasting are achieved by not assuming that assets are perfect substitutes. In contrast to Boughton, Fisher *et al.* also find that a failure to take into account the endogeneity of interest rates can lead to poor forecasting performance.

An alternative class of models adopted in the literature is based on the premise that poor forecasting power is not the result of a misspecification of the set of economic variables used to model the economic fundamentals governing the exchange rate, but a misspecification of the functional form. Meese and Rose (1991) argue the point strongly by showing that exchange rates exhibit significant nonlinearities. Unfortunately, in a fairly extensive empirical search they still cannot find a nonlinear model that dominates the random walk out-of-sample. Chinn (1991) provides some support for nonlinear exchange rate models, albeit weak evidence, by using the alternating conditional expectations model.

The class of models investigated by Meese and Rose (1991) and Chinn (1991) are nonparametric. More success has been achieved by using parametric nonlinear models. Engel and Hamilton (1990) use a Markovian switching model, Lye and Martin (1994) introduce the univariate generalised exponential nonlinear time series class GENTS, and in Chapter 8 Creedy *et al.* estimate a nonlinear error correction model. An important feature of the latter two models is that special attention is given to modelling the distribution of the exchange rate. This amounts to parameterising not only the first two moments of the distribution, as is the case with the class of linear models, but also higher-order moments such as skewness and kurtosis. The important property that drives the superior forecasting power of these models is that large movements in exchange rates are modelled arbitrarily well by conditional distributions that are bimodal, whereas small movements are modelled by univariate conditional distributions. Switching between univariate and bivariate distributions is determined endogenously within the model, and not *a priori* through inspection of the data.

This chapter investigates the forecasting performance of this new class of nonlinear parametric models and compares its forecasting performance with the existing class of linear models. In doing so, many of the forecasting models discussed in previous chapters are evaluated using a common data set. Both univariate and multivariate models are investigated. The univariate models consist of GENTS, Tong's (1983, 1990) SETAR model, a Markovian switching model, an artificial neural network (ANN) model as discussed in chapter 12, and a linear autoregressive model. The multivariate models include multivariate analogues of the GENTS and SETAR models, a structural model of exchange rates and a linear VAR. Section 14.1 summarises the competing models used in the empirical investigation. The properties of the data are discussed in Section 14.2 and the empirical results are reported in Section 14.3. Concluding comments are given in Section 14.4.

14.1 Nonlinear Forecasting Models

14.1.1 Univariate Models

GENTS

The generalised exponential nonlinear class of time series models, GENTS, introduced by Lye and Martin (1994) stems from the previous work on the generalized Student t distributions of Lye and Martin (1993a). The generalised Student t distribution is a subordinate of the generalised exponential class of distributions discussed by Cobb *et al.* (1983) and Martin (1990). An important property of the GENTS model is that it can capture a range of non-normal distributional shapes including skewness, leptokurtosis and multimodality.

Defining E_t as the nominal exchange rate, the GENTS(L) model specification adopted is based on the empirical work of Lye and Martin (1994):

$$E_t = \lambda_0 + \sum_{i=1}^{L} \lambda_i E_{t-i} + u_t \tag{14.1}$$

where L represents the length of the lag distribution and u_t is an error term which is distributed as generalised Student t:

$$f(u_t) = \exp\left[\Theta_{1,t} \tan^{-1}(u_t/\gamma) - \frac{1+\gamma^2}{2} \ln\left(\gamma^2 + u_t^2\right) + \Theta_{3,t} u_t - 0.5u_t^2 - \eta_t\right] \tag{14.2}$$

The term η_t, is the normalising constant which is given by:

$$\eta_t = \ln \int \exp\left[\Theta_{1,t} \tan^{-1}(u_t/\gamma) - \frac{1+\gamma^2}{2} \ln\left(\gamma^2 + u_t^2\right) + \Theta_{3,t} u_t - 0.5u_t^2\right] du_t \tag{14.3}$$

to ensure that the distribution integrates to unity. The $\Theta_{i,t}$ terms are assumed to have linear autoregressive representations:

$$\Theta_{1,t} = \theta_{1,0} + \sum_{i=1}^{L} \theta_{1,i} E_{t-i} \tag{14.4}$$

$$\Theta_{3,t} = \theta_{3,0} + \sum_{i=1}^{L} \theta_{3,i} E_{t-i} \tag{14.5}$$

The GENTS model in equations (14.1) to (14.5) shows that there are two linkages between E_t and lagged E_t. The first linkage is interpreted as a direct one and is given by the conditional mean in equation (14.1). The second is an indirect link which arises from the affect of lags in E_t on the distribution of E_t through the terms in (14.4) and (14.5). It is this second link which gives the model its intrinsically nonlinear features which arise even if (14.1) to (14.5) are specified as linear autoregressive representations; see also chapter 8.

The GENTS model is estimated by maximum likelihood methods, the details of which are contained in Lye and Martin (1994).

SETAR

The SETAR(L) model of Tong (1983, 1990) consists of piecewise linear autoregressive time series representations where switching between autoregressive processes can occur at each point in time. The model is:

$$E_t = \begin{cases} \alpha_0 + \sum\limits_{i=1}^{L} \alpha_i E_{t-i} + \upsilon_t; & E_{t-\tau} \leq \phi \\ \beta_0 + \sum\limits_{i}^{L} \beta_i E_{t-i} + \upsilon_t; & E_{t-\tau} > \phi \end{cases} \tag{14.6}$$

where τ is the delay parameter, ϕ is the threshold parameter and υ is an error term. It is convenient to rewrite the SETAR model as:

$$E_t = \beta_0 + (\alpha_0 - \beta_0) D_t + \sum_{i=1}^{L} (\alpha_i - \beta_i) E_{t-i} D_t + \sum_{i=1}^{L} \beta_i E_{t-i} + \upsilon_t \tag{14.7}$$

where D_t is a dummy variable which has the property:

$$D_t = \begin{cases} 1: & E_{t-\tau} \leq \phi \\ 0: & E_{t-\tau} > \phi \end{cases} \tag{14.8}$$

The SETAR model in (14.7) is estimated by OLS for given values of the threshold and delay parameters ϕ and τ, respectively. In turn, the threshold and delay parameters are chosen which minimize the sum of squared residuals.

Hamilton's Markovian Switching Model

A characteristic of the SETAR model is that the switch between models, as given by D_t in (14.8), is deterministic. An alternative model which allows the switch to be stochastic is the Markovian switching model of Engel and Hamilton (1990). Defining S_t as the binary stochastic switching variable with conditional probabilities:

$$\begin{aligned} p &= P[S_t = 1 | S_{t-1} = 1] \\ q &= P[S_t = 0 | S_{t-1} = 0] \end{aligned} \tag{14.9}$$

the exchange rate model is expressed as:

$$E_t = \alpha_0 + \alpha_1 S_t + \sqrt{\beta_0 + \beta_1 S_t} w_{1,t} \tag{14.10}$$

where $w_{1,t}$ is a $N(0,1)$ error term, and S_t has the autoregressive representation:

$$S_t = (1-q) + (p+q-1) S_{t-1} + \sqrt{q(1-q) + (p(1-p) - q(1-q)) S_t} w_{2,t} \tag{14.11}$$

where $w_{2,t}$ is a Bernoulli random variable. The parameters $\{p, q, \alpha_0, \alpha_1, \beta_0, \beta_1\}$ are estimated by maximum likelihood methods.[1]

[1] Pagan and Martin (1996) show how this model can be estimated using indirect estimation methods.

Artificial Neural Networks (ANN)

Following the framework developed in chapter 12, a single hidden layer artificial feedforward neural network with one activation function and L lags as inputs is given by:

$$E_t = \alpha_0 + \sum_{i=1}^{L} \alpha_i E_{t-i} + \frac{\phi}{1 + \exp\left[\gamma_0 + \sum_{i=1}^{L} \gamma_i E_{t-i}\right]} + \nu_t \tag{14.12}$$

where the parameters are $\{\alpha_i, \gamma_i, i = 0, 2, ..., L; \phi\}$, and ν_t is an error term. This model is denoted as ANN(L).[2] The parameters are estimated by choosing the γ_i parameters from a uniform distribution over the interval $(-2, 2)$, and then regressing E_t on a $\{$constant, $E_{t-1}, E_{t-2}, ..., E_{t-L}, Z_t\}$ where Z_t is the activation function given by:

$$Z_t = \frac{1}{1 + \exp\left[\widehat{\gamma}_0 + \sum_{i=1}^{L} \widehat{\gamma}_i E_{t-i}\right]} \tag{14.13}$$

and $\widehat{\gamma}_i \ \forall i$ represent the realisations from the random number generator. The final parameter estimates are obtained where the sum of squared errors is a minimum. The maximum number of iterations chosen is 100.[3]

AR

For comparison with the nonlinear models, the linear AR(L) model is given by:

$$E_t = \phi_0 + \sum_{i=1}^{L} \phi_i E_{t-i} + \varsigma_t \tag{14.14}$$

where ς_t is an error term. A special case of the AR model is the random walk where $\phi_0 = -, \phi_1 = 1, \phi_i = 0, i > 1$:

$$E_t = E_{t-1} + \varsigma_t \tag{14.15}$$

[2]It was noted in chapter 12 that a SETAR model can be interpreted as an ANN with a threshold function.

[3]See Kuan and White (1994) for an extension of this algorithm based on gradient methods.

14.1.2 Multivariate Models

Multivariate GENTS

The univariate GENTS model given by equations (14.1) to (14.5) is extended to a multivariate model by allowing for additional explanatory variables X_t, in the autoregressive representations (14.1), (14.4) and (14.5):

$$E_t = \lambda(E_{t-1}, E_{t-2}, ...; X_t) + u_t^M \tag{14.16}$$

$$\Theta_{i,t} = g_i(E_{t-1}, E_{t-2}, ...; X_t), \; i = 1, 3 \tag{14.17}$$

where u_t^M is an error term. Typical variables contained in X_t are the relative money monetary growth rates M_t, the relative growth rates in incomes Y_t, relative interest rate differentials R_t, and relative inflations P_t:

$$X_t = \{M_t, Y_t, R_t, P_t\} \tag{14.18}$$

To undertake out of sample forecasts of this model, AR(L) models for each of the variables in X_t are estimated and used to forecast the X_t variables out-of-sample.

Multivariate SETAR

A natural extension of the SETAR model is to consider multivariate versions whereby the model is augmented by the same set of explanatory variables X_t which are used to augment the GENTS model. The multivariate SETAR model is:

$$E_t = \beta_0 + (\alpha_0 - \beta_0) D_t + \sum_{i=1}^{L} (\alpha_i - \beta_i) E_{t-i} D_t + \sum_{i=1}^{L} \beta_i E_{t-i} + \sum_{i=1}^{4} \gamma_i X_{t,i} + \upsilon_t^M \tag{14.19}$$

where υ_t^M is an error term and D_t is defined in (14.8). As with the multivariate GENTS models, the $AR(L)$ models estimated for each of the variables in X_t are used to forecast the X_t variables out of sample.

Structural/Reduced Form

The structural model is based on the general reduced forms investigated by Boughton (1987) and Baillie and McMahon (1990). The difference is that the equations are expressed in returns and growth rates whereas previous structural models are expressed in levels. The structural equation is given by:

$$E_t = \beta_0 + \beta_1 M_t + \beta_2 Y_t + \beta_3 R_t + \beta_4 P_t + \omega_t \tag{14.20}$$

where ω_t is an error term and the explanatory variables are contained in (14.18). That is, M_t is the difference in the nominal money growth rates of the two countries, Y_t is the difference in the real income growth rates of the two countries, R_t is the difference in the interest rates of the two countries, and P_t is the difference in the inflation rates of the two countries.

As with the multivariate GENTS and SETAR models, the AR(L) models estimated for each of the variables in X_t are used to forecast the X_t variables out-of-sample.

VAR

A VAR(L) is a general linear dynamic model which allows for each variable to be expressed as a function of own lags and lags of all other variables in the model, where L denotes the number of lags. A five-variate VAR is estimated consisting of the variables $\{E_t, M_t, Y_t, R_t, P_t\}$ as well as a constant. The lag lengths are the same in each equation.

14.2 Statistical Properties of the Data

This section examines the properties of the data used in the forecasting experiments. Special attention is given to detecting evidence of nonlinearities using a range of tests and the close returns plots discussed in chapter 10.

14.2.1 Description

The exchange rate data (E_t) consist of monthly percentage returns on the Japanese/US bilateral nominal exchange rate beginning in February 1973 and ending in September 1993, a total of 248 observations. The variables used to capture market fundamental variables in the multivariate experiments are M_t differential in country nominal monetary growth rates; Y_t, differential in country real income growth rates; R_t, country nominal interest rate differential; and P_t, country inflation differential. All data are taken from the International Financial Statistics data base.

14.2.2 Tests of Nonlinearity

Table 14.1 contains a range of tests of nonlinearity of the Yen/US returns. All test results are based on the OLS residuals of estimating an AR(1) with a constant for the Yen/US foreign returns; ARCH denotes the test for first order ARCH; Neural(k) indicates the Lee *et al.* (1993) neural network test for hidden nonlinearity with k activation functions with all lag lengths equal

Table 14.1: Nonlinearity tests of the Yen/US returns

Test statistic	p-value
ARCH	0.1996
Skewness	0.1373
Kurtosis	0.0143
Jarque-Bera	0.0501
RESET	0.7502
Neural(1)	0.7421
Neural(2)	0.2728
Neural(3)	0.4344
Neural(4)	0.5817
BDS(2)	0.0289
BDS(3)	0.0018
BDS(4)	0.0005
BDS(5)	0.0003
BDS(6)	0.0001

to $L = 1$); BDS(m) denotes the Brock *et al.* (1986) test for nonlinearity using m-histories.

The first block of diagnostics are standard results and show no evidence of first order ARCH or skewness. The RESET test also shows no significant nonlinearities in the mean. The kurtosis test reveals evidence of fat tails in the error distribution at the 5 per cent level.

The neural network tests of hidden nonlinearities presented in Table 14.1 for activation functions ranging from 1 to 3, show no significant evidence of nonlinearities at the 5 per cent level. In contrast to these results, the BDS statistics using 2 to 6 m-histories inclusively, all show evidence of significant dependencies at the 5 per cent level.

14.2.3 Recurrence Plots

To investigate further the properties of the returns of the Yen/US exchange rate, the close returns plots discussed in detail in chapter 10 are presented in Figure 14.1 for embedding dimensions of $m = 1$ and $m = 4$. The close returns plot provides a graph of the index of reconstructed vectors against the sampling interval when vectors are at close returns. Processes which exhibit some structure, either linear or nonlinear, are identified by patterns. For chaotic series the close return plots are represented by horizontal lines where the length of the lines gives the period of the cycles. In the case of purely white noise processes, the close returns plot exhibits a random scatter of points thereby exhibiting no structure.

Inspection of the close returns plots in Figure 14.1 shows some evidence of clustering of points into diagonal line paths. This is suggestive of a nonrandom underlying structure which could be nonlinear. The possibility that the nonlinearities are the result of a chaotic process is ruled out as the close returns plots do not display horizontal line segments.

14.3 Empirical Comparisons of Models

The out-of-sample properties of alternative models for forecasting percentage monthly returns in the Japanese/US exchange rate are compared in this section. The maximum lag length chosen for all models is $L = 3$, thus the sample period effectively starts in May 1973 and the effective total number of observations is reduced to 245. The estimation period is from May 1973 to September 1992, leaving $H = 12$ observations to conduct out-of-sample forecasting tests.

14.3.1 Univariate Results

The results of the out-of-sample forecast performance of the various univariate models are given in Table 14.2. This table gives the ratio of the RMSE of a particular model to the RMSE obtained for a random walk model without drift, for various forecast horizons. A reported value less than unity is evidence that the random walk model yields inferior forecasts on average. The H-period ahead forecasts of Hamilton's stochastic switching model conditional on information at time T, are computed as:

$$F_{T+H} = \widehat{\alpha}_0 + \left(\widehat{\rho} + (-1 + \widehat{p} + \widehat{q})^H \left(P\left[S_t = 1 | E_1, E_2, ..., E_T; \widehat{\theta}\right] - \widehat{\rho}\right)\right)\widehat{\alpha}_1$$

where $\widehat{\theta} = \{\widehat{p}, \widehat{q}, \widehat{\alpha}_0, \widehat{\alpha}_1, \widehat{\beta}_0, \widehat{\beta}_1\}$ is the set of maximum likelihood parameter estimates based on the sample period, $P\left[S_t = 1 | E_1, E_2, ..., E_T; \widehat{\theta}\right]$ is obtained from the likelihood and:

$$\widehat{\rho} = \frac{1 - \widehat{q}}{2 - \widehat{p} - \widehat{q}}$$

Inspection of Table 14.2 shows that the GENTS model with up to two lags is the only model which consistently beats the linear AR model over all forecast horizons. For the 1-month and 4-month forecast horizons, GENTS(1) and GENTS(2) perform the best, followed by the AR models and Hamilton's

Table 14.2: Forecasting performance of univariate exchange rate models[a]

Model	Forecast period: M (mths)			
	1	4	8	12
GENTS(1)	1.0607	1.0831	0.9169	0.9295
GENTS(2)	0.8409	0.8224	0.9288	0.9409
GENTS(3)	1.3014	1.2558	1.0453	1.0458
SETAR(1)	1.2363	1.3605	0.9057	0.9171
SETAR(2)	1.2773	1.4111	0.9127	0.9246
SETAR(3)	1.1982	1.4138	0.9061	0.9190
Hamilton	1.1787	1.2364	0.9420	0.9434
ANN(1)	1.2479	1.3527	0.9135	0.9227
ANN(2)	1.2525	1.4088	0.8939	0.9111
ANN(3)	1.3525	1.5754	0.8750	0.9145
AR(1)	1.1579	1.1808	0.9674	0.9666
AR(2)	1.1772	1.2044	0.9715	0.9705
AR(3)	1.1211	1.1971	0.9677	0.9669

(a) Ratio of model RMSE to random walk RMSE.

model. The worst performers are the SETAR and ANN models. For the 1-year forecast horizon, all nonlinear models display superior forecasting power to the linear AR models.

14.3.2 Multivariate Results

The forecast results of Yen/US returns from various multivariate models are given in Table 14.3. Comparing these results with the univariate results in Table 14.2 shows that there is little information contained in the economic variables $\{M_t, Y_t, R_t, P_t\}$ as the RMSEs for the univariate models tend to be lower than the RMSEs for the multivariate models for a range of models specifications and for a range of forecast horizons.

14.4 Conclusions

This chapter examined a range of models, including linear (AR and VAR) and nonlinear univariate and multivariate models, for predicting monthly US/Yen returns. The nonlinear models considered were the GENTS and SETAR nonlinear time series models, Hamilton's switching model and an

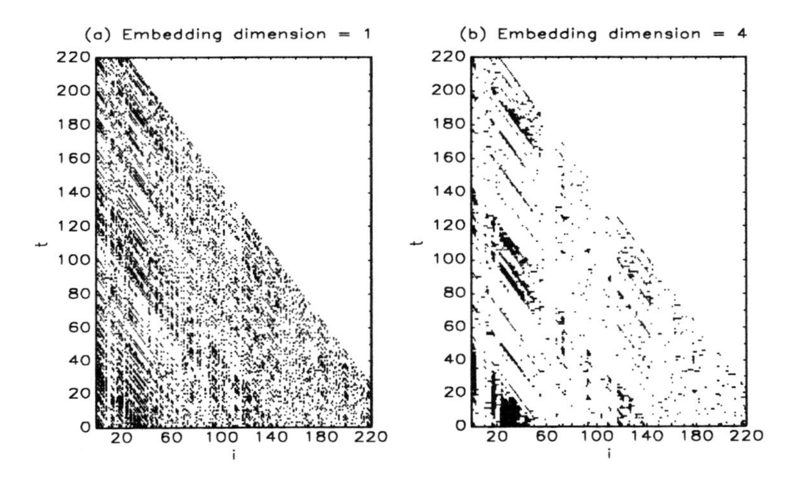

Figure 14.1: Close returns plots of Japanese/US exchange rate returns

Table 14.3: Forecasting performance of multivariate exchange rate models[a]

Model	Forecast period: M (months)			
	1	4	8	12
MGENTS(1)	1.0421	1.0626	0.9056	0.9390
MGENTS(2)	1.0727	1.0839	1.0433	1.0659
MGENTS(3)	1.3846	1.3585	1.0872	1.0909
MSETAR(1)	1.3287	1.4204	0.9328	0.9418
MSETAR(2)	1.3624	1.4637	0.9393	0.9489
MSETAR(3)	1.2745	1.4529	0.9292	0.9396
VAR(1)	1.2714	1.3055	0.9586	0.9590
VAR(2)	1.4859	1.4954	0.9960	0.9935
VAR(3)	1.3929	1.3616	0.9502	0.9510

(a) Ratio of model RMSE to random walk RMSE.

ANN model. Out-of-sample forecast comparisons were made for 1,4, 8 and 12 month forecast horizons.

The key result of the forecast experiments is that the nonlinear time series model, GENTS, performs the best of all models across all forecast horizons. The GENTS class of models is the only class to dominate the linear models consistently. The strength of the GENTS models was strongest for the 1 and 4 month horizons whereas the neural network models followed by the SETAR models were the worst. For the 1-year forecast horizon, all nonlinear models produced similar forecasting power, and were superior to the linear models. The results also showed that there were no gains in forecasting, over one year at least, from including additional variables to capture the market fundamentals.

Bibliography

[1] Aitchison, J. and Brown, J.A.C. (1957) *The Lognormal Distribution.* London: Cambridge University Press.

[2] Akgiray, V. and Booth, G.G. (1988) Mixed Diffusion-Jump Process Modelling of Foreign Exchange Rate Movements. *Review of Economics and Statistics*, 770, 631-637.

[3] Allen, F. and Karjalainen, R. (1994) Using Genetic Algorithms to Find Technical Trading Rules. *The Wharton School University of Pennsylvania Working Paper.*

[4] Amemiya, T. (1981) Qualitative Response Models: A Survey. *Journal of Economic Literature*, XIX, 1483-1536.

[5] Amemiya, T. and Boskin, M. (1974) Regression Analysis When the Dependent Variable is Truncated Lognormal, with an Application to the Determinants of the Duration of Welfare Dependency. *International Economic Review*, 15, 485-496.

[6] Andrews, D.W.K. (1993) Tests for Parameter Instability and Structural Change with Unknown Change Point. *Econometrica*, 61, 821-856.

[7] Andrews, D.W.K. and Ploberger, W. (1994) Optimal Tests When a Nuisance Parameter Is Present Only Under the Alternative. *Econometrica*, 62, 1383-1414.

[8] Backus, D. (1984) Empirical Models of the Exchange Rate: Separating the Wheat from the Chaff. *Canadian Journal of Economics*, XVII, 824-846.

[9] Baillie, R. and McMahon, P. (1990) *The Foreign Exchange Market: Theory and Econometric Evidence.* Cambridge: Cambridge University Press,

[10] Baillie, R.T. and Pecchenino, R.A. (1991) The Search for Equilibrium Relationships in International Finance: The Case of the Monetary Model. *Journal of International Money and Finance*, 10, 582-593.

[11] Bakker A. (1995a) Extensions of the Generalized Least Squares Frequency Estimator. *University of Melbourne Department of Economics Research Paper,* 499.

[12] Bakker, A. (1995b) The Least Squares Frequency Estimator of the Generalized Exponential Family of Densities: Inference and Monte Carlo Evidence. *University of Melbourne Department of Economics Research Paper,* 491.

[13] Barsky, R.B. and De Long, J.B. (1993) Why Does the Stock Market Fluctuate? *Quarterly Journal of Economics,* CVIII, 291-311.

[14] Bauer, R.J. (1994) *Genetic Algorithms and Investment Strategies.* New York: John Wiley.

[15] Beard, R.E., Pentikäinen, T. and Pesonen, E. (1984) *Risk Theory,* 3rd edition, London: Chapman and Hall.

[16] Bera, A., Jarque, C. and Lee, L.F. (1984) Testing the Normality Assumption in Limited Dependent Variable Models. *International Economic Review,* 25, 563-578.

[17] Berndt, E.R., Hall, B., Hall, R.E. and Hausman, J.A. (1974) Estimation and Inference in Non-linear Structural Models. *Annals of Economic and Social Measurement,* 3, 653-665.

[18] Berndt, E.R. (1991) *The Practice of Econometrics: Classic and Contemporary.* New York: Addison-Wesley.

[19] Bishop, Y.M.M. (1975) *Discrete Multivariate Analysis: Theory and Practice.* Cambridge, Mass: The MIT Press.

[20] Bollerslev, T. (1986) Generalized Autoregressive Conditional Heteroscedasticity. *Journal of Econometrics,* 21, 307-328.

[21] Boothe, P. and Glassman, D. (1987) The Statistical Distribution of Exchange Rates: Empirical Evidence and Economic Implications. *Journal of International Economics,* 22, 297-319.

[22] Boughton, J.M. (1987) Tests of the Performance of Reduced-form Exchange Rate Models. *Journal of International Economics,* 23, 41-56.

[23] Brock, W.A. and Malliaris, A.G. (1982) *Stochastic Methods in Economics and Finance.* New York: North-Holland.

[24] Brock, W.A. and Malliaris, A.G. (1989) *Differential Equations, Stability and Chaos in Dynamic Economics.* Amsterdam: North-Holland-Elsevier.

[25] Brock, W.A. and Sayers, C.L. (1988) Is the Business Cycle Characterized by Deterministic Chaos? *Journal of Monetary Economics,* 22, 71-90.

[26] Brock, W.A., Dechert, W.D. and Scheinkman, J. A. (1987) A Test for Independence Based on the Correlation Dimension. *University of Wisconsin, Madison.*

[27] Brock, W., Lakonishok, J. and LeBaron, B. (1992) Simple Technical Trading Rules and the Stochastic Properties of Stock Returns. *Journal of Finance*, 47, 1731-1764.

[28] Brown, J.A.C. (1976) The Mathematical and Statistical Theory of Income Distribution. In *The Personal Distribution of Income* (ed. A. B. Atkinson), 72-97. London: George Allen & Unwin.

[29] Champernowne, D.G. (1953) A Model of Income Distribution. *Economic Journal*, 63, 316-351.

[30] Chao, J.C. and Phillips, P.C.B. (1996) Bayesian Posterior Distributions in Limited Information Analysis of the Simultaneous Equations Model. *Yale University, Cowles Foundation.*

[31] Chinn, M.D. (1991) Some Linear and Nonlinear Thoughts on Exchange Rates. *Journal of International Money and Finance*, 10, 214-230.

[32] Cobb, L., and Zacks, S. (1985) Applications of Catastrophe Theory for Statistical Modelling in the Biosciences. *Journal of the American Statistical Association*, 80, 793-802.

[33] Cobb, L., Koppstein, P. and Chen, N.H. (1983) Estimation and Moment Recursion Relations for Multimodal Distributions of the Exponential Family. *Journal of the American Statistical Association*, 78, 124-130.

[34] Copenhaver, T.W. and Mielke, P.W. (1977) Quantit Analysis: Quantal Assay Refinement. *Biometrics,* 33, 175-186.

[35] Cosslett, S.R. (1983) Distribution Free Maximimum Likelihood Estimator of the Binary Choice Model. *Econometrica,* 51, 765-782.

[36] Cox, D.R. and Miller, H.D. (1984) *The Theory of Stochastic Processes.* London: Chapman & Hall.

[37] Creedy, J. (1985) *Dynamics of Income Distribution.* Oxford: Basil Blackwell.

[38] Creedy, J. (1992) *Demand and Exchange in Economic Analysis.* Aldershot: Edward Elgar.

[39] Creedy, J. (1996) *General Equilibrium and Welfare.* Aldershot: Edward Elgar.

[40] Creedy, J. and Martin, V.L. (1993) Multiple Equilibria and Hysteresis in Simple Exchange Models. *Economic Modelling*, 10, 339-347.

[41] Creedy, J. and Martin, V.L. (1994a) The Strange Attraction of Chaos in Economics. In *Chaos and Nonlinear Models in Economics: Theory and Applications* (eds. J. Creedy and V.L. Martin). Aldershot: Edward Elgar.

[42] Creedy, J. and Martin, V.L. (1994b) A Model of the Distribution of Prices. *Oxford Bulletin of Economics and Statistics*, 56, 67-76.

[43] Creedy, J. and Martin, V.L. (eds) (1994c) *Chaos and Non-linear Models in Economics*. Aldershot: Edward Elgar.

[44] Creedy, J., Lye, J. and Martin, V.L. (1994) Non-linearities and the Long-run Real Exchange Rate Distribution. In *Chaos and Nonlinear Models in Economics: Theory and Applications* (eds. J. Creedy and V.L. Martin). Aldershot: Edward Elgar.

[45] Creedy, J., Lye, J. and Martin, V.L. (1996a) A Labour Market Model of Income Distribution. *Journal of Income Distribution,* 6, 127-144.

[46] Creedy, J., Lye, J.N. and Martin, V.L. (1996b) A Nonlinear Model of the Real US/UK Exchange Rate. *Journal of Applied Econometrics*, 11, 669-686.

[47] Deboeck, G.J.(1994) *Trading on the Edge: Neural, Genetic, and Fuzzy Systems for Chaotic Financial Markets.* New York: John Wiley.

[48] Demuth, H. and Beale, M. (1994) *Neural Network Toolbox for Use with MATLAB.* Natick: The MathWorks Inc.

[49] Diebold, F.X. (1988) *Empirical Modelling of Exchange Rate Dynamics.* Berlin: Springer-Verlag.

[50] Diebold, F.X. and Nason, J.M. (1990) Nonparametric Exchange Rate Prediction. *Journal of International Economics*, 28, 315-332.

[51] Domenich, T.A. and McFadden, D. (1975) *Urban Travel Demand.* Amsterdam: North-Holland.

[52] Domowitz, I. and Hakkio, C.S. (1985) Conditional Variance and the Risk Premium in the Foreign Exchange Market. *Journal of International Economics*, 19, 47-66.

[53] Dorsey, R.E. and Mayer, W.J. (1995) Genetic Algorithms for Estimation Problems with Multiple Optima, Nondifferentiability, and Other Irregular Features. *Journal of Business and Economic Statistics*, 13(1), 53-66.

[54] Eckmann, J. P., Kamphorst, S. O. and Ruelle, D. (1987) Recurrence Plots of Dynamical System. *Europhysics Letters*, 4, 973-977.

[55] Eng, W.F. (1988) *The Technical Analysis of Stocks, Options, and Futures: Advanced Trading Systems and Techniques.* Chicago: Probus.

[56] Engel, C. and Hamilton, J.D. (1990) Long Swings in the Dollar: Are They in the Data and Do Markets Know It? *American Economic Review*, 80, 689-713.

[57] Engle, R. F. (1982) Autoregressive Conditional Heteroscedasticity with estimates of the Variance of United Kingdom Inflation. *Econometrica*, 50, 987-1007.

[58] Engle, R.F. and Granger, C.W.J. (1987) Co-integration and Error Correction: Representation, Estimation and Testing. *Econometrica,* 55, 251-276.

[59] Finch P.D. (1989) An Approximation to ML-Estimation. *Monash University Department of Statistics.*

[60] Fischer, E.O. and Jammernegg, W. (1986) Empirical Investigation of a Catastrophe Theory Extension of the Phillips Curve. *Review of Economics and Statistics*, 68, 9-17.

[61] Fisher, P.G., Tanna, S.K., Turner, D.S., Wallis, K.F. and Whitley, J.D. (1990). *Economic Journal*, 100, 1230-1244.

[62] Flood, R.P. and Garber, P.M. (1980) Market Fundamentals Versus Price Level Bubbles: The First Tests. *Journal of Political Economy*, 89, 745-770.

[63] Frank, M.Z. and Stengos, T. (1988) Some Evidence Concerning Macroeconomic Chaos. *Journal of Monetary Economics*, 22, 423-438.

[64] Frank, M.Z. and Stengos, T. (1989) Measuring the Strangeness of Gold and Silver Rates of Return. *Review of Economic Studies*, 56, 553-567.

[65] Frankel, J.A. (1991) Is a Yen Bloc Forming in Pacific Asia? In *Finance and the International Economy* (ed. R. O'Brien). New York: Oxford University Press.

[66] Frankel, J.A. and Wei, S.J. (1993) Is There a Currency Bloc in the Pacific? In *The Exchange Rate, International Trade and the Balance of Payments* (ed. A. Blundell-Wignall). Sydney: Reserve Bank of Australia.

[67] Friedman, D. and Vandersteel, S. (1982) Short-run Fluctuations in Foreign Exchange Rates: Evidence from the Data 1973-79. *Journal of International Economics*, 13, 171-186.

[68] Fuller, W.A. (1976) *Introduction to Statistical Time Series.* New York: John Wiley.

[69] Gabler, S., Laisney F. and Lechner, M. (1993) Seminonparametric Estimation of Binary Choice Models With an Application to Labor-force Participation. *Journal of Business and Economic Statistics*, 11, 1, 61-80.

[70] Gallant, A.R. and Nychka, D.N. (1987) Semi-nonparametric Maximum Likelihood Estimation. *Econometrica*, 55, 363-390.

[71] Gallant, A.R., Rossi, P.E. and Tauchen,G. (1992) Stock Prices and Volume. *Review of Financial Studies*, 5, 199-242.

[72] Gennotte, G. and Leland, H. (1990) Market Liquidity, Hedging, and Crashes. *American Economic Review*, 80, 999-1021.

[73] Gilmore, C.G. (1993) A New Test for Chaos. *Journal of Economic Behavior and Organization*, 22, 209-237.

[74] Gilmore, R. (1981) *Catastrophe Theory for Scientists and Engineers.* New York: John Wiley.

[75] Goldberg, D.E. (1989) *Genetic Algorithms in Search, Optimisation and Machine Learning.* Reading, MA: Addison-Wesley.

[76] Gourieroux, C., Monfort, A. and Renault, E. (1993) Indirect Inference. *Journal of Applied Econometrics*, 8, S85-S118.

[77] Grassberger, P. and Proccacia, I. (1983a) Measuring the Strangeness of Strange Attractors. *Physica D*, 90, 189-208.

[78] Grassberger, P. and Proccacia, I. (1983b) Characterization of Strange Attractors. *Physical Review Letters*, 50, 346-349.

[79] Greene, W.H. (1990) *Econometric Analysis.* New York: Macmillan.

[80] Gregory, A.W. and Hansen, B.E. (1996) Residual-Based Tests for Cointegration in Models with Regime Shifts. *Journal of Econometrics*, 70, 99-126.

[81] Guckenheimer, J. (1982) Noise in Chaotic Systems. *Nature*, 298, 358-361.

[82] Hackl, P. (1989) *Statistical Analysis and Forecasting of Economic Structural Change.* Berlin: Springer-Verlag.

[83] Hart, P.E. (1973) Random Processes and Economic Size Distributions. *University of Reading.*

[84] Harvey, A.C. (1990) *The Econometric Analysis of Time Series.* New York: Philip Allan.

[85] Hausman, J. and Wise, D. (1977) Social Experimentation, Truncated Distributions and Efficient Estimation: *Econometrica*, 45, 919-938.

[86] Heckman, J.J. and Willis, R.J. (1977) A Beta-logistic Model for the Analysis of Sequential Labor Force Participation by Married Women. *Journal of Political Economy*, 85, 27-58.

[87] Hesselager, O. (1994) A Recursive Procedure for Calculation of Some Compound Distributions, *ASTIN Bulletin*, 24, 19-32.

[88] Hogg, R.V. and Klugman, S.A. (1983) On the Estimation of Long-tailed Skewed Distributions with Actuarial Applications. *Journal of Econometrics*, 23, 91-102.

[89] Holland, J.H. (1975) *Adaption in Natural and Artificial Systems.* Ann Arbor, MI: University of Michigan Press.

[90] Hsieh, D.A. (1989a) Testing for Nonlinear Dependence in Daily Foreign Exchange Rates. *Journal of Business*, 62, 339-368.

[91] Hsieh, D.A. (1989b) Modeling Heteroskedasticity in Daily Foreign Exchange Rates. *Journal of Business and Economic Statistics*, 7, 307-317.

[92] Johnson, N.L. and Kotz, S. (1970) *Continuous Univariate Distributions.* New York: John Wiley.

[93] Kamien, M.I. and Schwartz, N.L. (1981) *Dynamic Optimization: The Calculus of Variations and Optimal Control in Economics and Management.* New York: North-Holland.

[94] Klein, L. R. (1962) *An Introduction to Econometrics.* Englewood Cliffs, NJ: Prentice-Hall.

[95] Klein, R.W. and Spady, R.H. (1987) An Efficient Semiparametric Estimator for Discrete Choice Models. Bellcore, *Discussion paper.*

[96] Kleibergen, F. and van Dijk, H.K. (1994) On the Shape of the Likelihood/Posterior in Cointegration Models. *Econometric Theory*, 10, 514-551.

[97] Kleibergen, F. and van Dijk, H.K. (1996) Bayesian Simultaneous Equations Analysis Using Reduced Rank Structures. *Rotterdam University, Econometric Institute and Tinbergen Institute Working Paper.*

[98] Koebbe, M. and Mayer-Kress, G. (1992) Use of Recurrence Plots in the Analysis of Time Series Data. *Department of Mathematics, University of Santa Cruz*

[99] Krommer, A.R. and Ueberhuber, C.W. (1994) *Lecture Notes in Computer Science: Numerical Integration on Advanced Computer Systems.* Berlin, New York: Springer-Verlag.

[100] Kuan, C.M. and White, H. (1994) Artificial Neural Networks: An Econometric Perspective. *Econometric Reviews*, 13, 1-91.

[101] LeBaron, B. (1992) Do Moving Average Trading Rule Results Imply Nonlinearities in Foreign Exchange Markets? *University of Wisconsin-Madison, Social Systems Research Institute. Working Paper 9222*

[102] Le Baron, B. (1994) Technical Trading Rule Profitability and Foreign Exchange Intervention. *University of Wisconsin, Madison, Technical Report.*

[103] LeBaron, B. and Scheinkman, J. A. (1987) Non-linear Dynamics and GNP Data. *University of Chicago, Department of Economics.*

[104] Lee, T., White, H. and Granger C.W.J. (1993) Testing for Neglected Nonlinearity in Time Series Models: A Comparison of Neural Network Methods and Alternative Tests. *Journal of Econometrics*, 56, 269-290.

[105] Lim, G.C. and Martin, V.L. (1995) Regression-based Cointegration Estimators with Applications. *Journal of Economic Studies*, 22, 3-22.

[106] Lim, G.C., Martin, V.L. and Teo, L. (1995) A Nonlinear Characterization of Asset Price Dynamics with an Application to the 1987 Stock Market Crash. *University of Melbourne, Department of Economics Research Paper.*

[107] Lim, G.C., Lye, J.N., Martin, G.M. and Martin, V.L. (1996) The Distribution of Exchange Rate Returns: An International Comparison. *University of Melbourne, Department of Economics Research Paper.*

[108] Lukac, L.P. and Brorsen, B.W. (1989) The Usefulness of Historical Data in Selecting Parameters for Technical Trading Systems. *Journal of Futures Markets*, 9, 55-65.

[109] Lydall, H. F. (1976) Theories of the Distribution of Earnings. In *The Personal Distribution of Income* (ed. A. B. Atkinson), 15-46. London: George Allen and Unwin.

[110] Lydall, H. F. (1979) *A Theory of Income Distribution.* Oxford: Clarendon Press.

[111] Lye, J.N. and Martin, V.L. (1993a) Robust Estimation, Nonnormalities and Generalized Exponential Distributions. *Journal of the American Statistical Association*, 88, 253-259.

[112] Lye, J.N. and Martin, V.L. (1993b) A Flexible Parametric Density Estimator for Multimodal Distributions of Test Statistics. *Communications in Statistics*, 22, 813-830.

[113] Lye, J.N. and Martin, V.L. (1994) Non-linear Time Series Modelling and Distributional Flexibility. *Journal of Time Series Analysis*, 15, 65-84.

[114] Maddala, G.S. (1983) *Limited-dependent and Qualitative Variables in Econometrics.* Cambridge: Cambridge University Press.

[115] Malliaris, A.G. and Brock, W.A. (1982) *Stochastic Methods in Economics and Finance.* New York: North-Holland.

[116] Mandelbrot, B.B. (1963) The Variation of Certain Speculative Prices. *Journal of Business*, 36, 394-419.

[117] Mandelbrot, V. (1960) The Pareto-Levy Law and the Distribution of Income. *International Economic Review*, 1, 79-106.

[118] Marron, J.S. and Schmitz, H.P. (1992) Simultaneous Estimation of Several Income Distributions. *Econometric Theory*, 8, 476-488.

[119] Martin, G.M. (1996) *Bayesian Inference in Models of Cointegration: Methods and Applications.* Ph.D. thesis (Monash University).

[120] Martin, V.L. (1990) *Properties and Applications of Distributions from the Generalized Exponential Family.* Ph.D. thesis (Monash University).

[121] McCaffrey, D., Ellner, S., Gallant, A.R. and Nychka, D. (1992) Estimating the Lyapunov Exponent of a Chaotic System with Nonparametric Regression. *Journal of the American Statistical Association*, 87, 682-695.

[122] McCulloch, W.S. and Pitts, W. (1943) A Logical Calculus of the Ideas Immanent in Nervous Activity. *Bulletin of Mathematical Biophysics.* 5, 115-133.

[123] McDonald, J.B. (1984) Some Generalized Functions for the Size Distribution of Income. *Econometrica*, 52, 647-663.

[124] Medsker, L., Turbin, E. and Trippi, R. (1993) Neural Network Fundamentals for Financial Analysts. In *Neural Networks in Finance and Investing* (eds. E. Turbin and R. Trippi). Chicago: Probus Publishing

[125] Meese, R.A. and Rogoff, K. (1983) Empirical Exchange Rate Models of the Seventies: Do They Fit Out of Sample? *Journal of International Economics*, 14, 3-24..

[126] Meese, R.A. and Rose, A.K. (1991) An Empirical Assessment of Nonlinearities in Models of Exchange Rate Determination. *Review of Economic Studies*, 58, 603-619.

[127] Mincer, J. (1970) Distribution and Labour Incomes: A Survey. *Journal of Economic Literature*, 8, 1-26.

[128] Mindlin, G.B., Solari, H.G., Natiello, M.A. Gilmore, R., Hou, X.J. and Tufillaro, N.B. (1990) Classification of Strange Attractors by Integers. *Physical Review Letters* , 64, 2350-2353.

[129] Mindlin, G.B., Solari, H.G., Natiello, M.A., Gilmore, R. and Hou, X.-J. (1991) Topological Analysis of Chaotic Time Series Data from the Belousov-Zhabotinskii Reaction. *Journal of Nonlinear Science*, 1, 147-173.

[130] Nadaraya, E.A. (1964) On Estimating Regression. *Theory of Probability and Its Applications*, 9, 141-142.

[131] Neftci S.N. (1984) Are Economic Time Series Asymmetric Over the Business Cycle? *Journal of Political Economy*, 92, 307-328

[132] Pagan, A.R. and Martin, V.L. (1996) Simulation Based Estimation of Some Factor Models in Econometrics. In *Simulation Based Inference in Econometrics: Methods and Applications* (eds. R.S. Mariano, M. Wrecks, and T. Schuermann). Cambridge: Cambridge University Press.

[133] Pagan, A.R. and Ullah , A. (1996) *Non-parametric Econometrics*, mimeo.

[134] Pagan, A.R. and Vella, F. (1989) Diagnostic Tests for Models Based on Individual data: A Survey. *Journal of Applied Econometrics*, 4, 29-59.

[135] Panjer, H.H. and Willmot, G.E. (1992) *Insurance Risk Models.* Schaumberg, IL: Society of Actuaries.

[136] Perron, P. and Vogelsang, T.J. (1992) Nonstationarity and Level Shifts with an Application to Purchasing Power Parity. *Journal of Business and Economic Statistics*, 10, 301-320.

[137] Pesaran, M.H. and Samiei, H. (1992) Estimating Limited-dependent Rational Expectations Models with an Application to Exchange Rate Determination in a Target Zone. *Journal of Econometrics*, 53, 141-163.

[138] Pesaran, M.H. and Shin, Y. (1994) Cointegration and Speed of Convergence to Equilibrium. University of Cambridge.

[139] Phelps Brown, E.H. (1977) *The Inequality of Pay.* Oxford: Oxford University Press.

[140] Phillips, P.C.B. (1987) Time Series Regressions with a Unit Root. *Econometrica*, 55, 277-302.

[141] Phillips, P.C.B. (1991a) To Criticize the Critics: An Objective Bayesian Analysis of Stochastic Trends. *Journal of Applied Econometrics*, 6, 333-364.

[142] Phillips, P.C.B. (1991b) Bayesian Routes and Unit Roots: de rebus prioribus semper est disputandum. *Journal of Applied Econometrics*, 6, 435-474.

[143] Phillips, P.C.B. (1991c) Error Correction and Long-run Equilibrium in Continuous Time. *Econometrica*, 59, 967-980.

[144] Phillips, P.C.B. (1993) The long-run Australian consumption function re-examined: an empirical exercise in Bayesian inference. *Yale University Cowles Foundation Paper No. 825.*

[145] Phillips, P.C.B. and Loretan, M. (1991) Estimating Long Run Economic Equilibria. *Review of Economic Studies* 58, 407-436.

[146] Pictet O.V., Dacorogna M.M., Muller U.A., Olsen R.B., and Ward J.R. (1992) Real-time Trading Models for Foreign Exchange Rates. *Neural Network World*, 2, 713-744.

[147] Poirier, D. J. (1978) The Use of the Box-Cox Transformation in Limited Dependent Variable Models. *Journal of the American Statistical Association*, 73, 284-287.

[148] Powell, J.L. (1983) Asymptotic Normality of the Censored and Truncated Least Absolute Deviations Estimators. *Stanford University, Technical Report 395, Institute for Mathematical Studies in the Social Sciences.*

[149] Prentice, R.L. (1976) A Generalization of the Probit and Logit Methods for Dose Response Curves. *Biometrics*, 32, 761-768.

[150] Pruitt, S.W., Tse, K.M. and White, R.E. (1992) The CRISMA Trading System: The Next Five Years. *Journal of Portfolio Management*, 22-25.

[151] Quandt, R.E. (1966) Old and New Methods of Estimation and the Pareto Distribution. *Metroeconomica*, 10, 55-82.

[152] Ramsey, J.B. and Yuan, H.J. (1989) Bias and Error Bias in Dimension Calculation and their Evaluation in Some Simple Models. *Physical Letters A*, 134, 287-297.

[153] Ramsey, J.B. and Yuan, H.J. (1990) The Statistical Properties of Dimension Calculations Using Small Data Sets. *Nonlinearity*, 3, 155-175.

[154] Robbins, L. (1930) On the Elasticity of Demand for Income in Terms of Effort. *Economica*, 10, 123-129.

[155] Ruohonen, M. (1988) A Model for the Claim Number Process. *ASTIN Bulletin*, 18, 57-68.

[156] Ruud, P.A. (1983) Sufficient Conditions for the Consistency of Maximum Likelihood Estimation Despite Misspecification of Distribution. *Econometrica*, 51, 225-228.

[157] Ruud, P.A. (1986) Consistent Estimation of Limited Dependent Variable Models Despite Misspecification of Distribution. *Journal of Econometrics*, 32, 157-187.

[158] Saikkonen, P. (1991) Asymptotically Efficient Estimation of Cointegrating Regressions. *Econometric Theory*, 7, 1-21.

[159] Salem, A.Z.B. and Mount, T.D. (1974) A Convenient Descriptive Model of Income Distribution: The Gamma Density. *Econometrica*, 42, 1115-1127.

[160] Sarle, W.S. (1994) Neural Networks and Statistical Models. *Proceedings of the Nineteenth Annual SAS Users Group International Conference*, NC: SAS Institute, 1538-1550.

[161] Scheinkman, J. and Le Baron, B. (1989) Nonlinear Dynamics and Stock Returns. *Journal of Business*, 62, 311-337.

[162] Schinasi, G. and Swamy, P.A.V.B. (1989) The Out-of-sample Forecasting Performance of Exchange Rate Models when Coefficients are Allowed to Change. *Journal of International Money and Finance*, 8, 375-390.

[163] Shorrocks, A.F. (1975) On Stochastic Models of Size Distribution. *Review of Economic Studies*, 42, 631-641.

[164] Simon, L. (1961) Fitting Negative Binomial Distributions by the Method of Maximum Likelihood. *Proceedings of the Casualty Actuarial Society*, 48, 45-53.

[165] Soong, T.T. (1973) *Random Differential Equations in Science and Engineering*. New York: Academic Press.

[166] Sundt, B. (1992) On Some Extensions of Panjer's Class of Distributions. *ASTIN Bulletin*, 22, 61-80.

[167] Taylor, S.J. (1986) *Modelling Financial Time Series*. New York: John Wiley.

[168] Teräsvirta, T. and Anderson, H.M. (1992) Characterizing Nonlinearities in Business Cycles Using Smooth Transition Autoregressive Models. *Journal of Applied Econometrics*, 7, S119-S136.

[169] Thyrion, P. (1961) Contribution à l'Étude du Bonus pour Non Sinistre en Assurance Automobile. *ASTIN Bulletin*, 1, 142-162.

[170] Tong, H. (1983) *Threshold Models in Non-linear Time Series Analysis*. Berlin: Springer-Verlag.

[171] Tong, H. (1990) *Non-linear Time Series: A Dynamical System Approach*. Oxford: Oxford University Press.

[172] Trippi, R.R. and Turban, E. (eds) (1996) *Neural Networks in Finance and Investing*. Chicago: Irwin, Chapt. 15, 329-365.

[173] Tucker, A.L. and Pond, L. (1988) The Probability Distribution of Foreign Exchange Price Changes: Tests of Candidate Processes. *The Review of Economics and Statistics*, 70, 638-647.

[174] Venables, A.J. (1984) Multiple Equilibria in the Theory of International Trade with Monopolistically Competitive Commodities. *Journal of International Economics*, 16, 103-121.

[175] Watson, G.S. (1964) Smooth Regression Analysis. *Sankhya*, A, 26, 359-372.

[176] Weigend, A.S., Huberman B.A., and Rumelhart, D.E. (1992) Predicting Sunspots and Exchange Rates with Connectionist Networks. In *Nonlinear Modeling and Forecasting* (eds. M. Casdagli and S. Eubank). 395-432. Redwood, California: Addison-Wesley.

[177] White, H. (1982) Maximum Likelihood Estimation of Misspecified Models. *Econometrica*, 50, 1-25.

[178] White, H. (1988) Econometric Predictions Using Neural Networks: The case of IBM Stock Prices. In *Proceedings of the Second Annual IEEE Conference in Neural Networks*, II: 451-458. New York: IEEE Press.

[179] White, H. (1989) Some Asymptotic Results for Learning in Single Hidden Layer Feedforward Network Models. *Journal of the American Statistical Association,* 84, 1003-1013.

[180] White, H. (1992) *Artificial Neural Networks: Approximation and Learning Theory.* Oxford: Basil Blackwell.

[181] Whitley, D. (1989) The GENITOR Algorithm and Selection Pressure: Why Rank-based Allocation of Reproductive Trials is Best. In *Proceedings of the Third International Conference on Genetic Algorithms* (ed. D.J. Schaffer), pp.116-121. San Mateo, California: Morgan Kaufmann.

[182] Willmot, G.E. (1993) On Recursive Evaluation of Mixed Poisson Probabilities and Related Quantities. *Scandinavian Actuarial Journal*, 2, 114-133.

[183] Wolf, A., Swift, J. B., Swinney, H. L. and Vastano, J. A. (1985) Determining Lyapunov Exponents From a Time Series. *Physica D*, 285-317.

Index